T0076430

THE IMPORTANCE OF BEING EDUCABLE

The Importance
of
Being Educable

A NEW THEORY OF
HUMAN UNIQUENESS

Leslie Valiant

PRINCETON UNIVERSITY PRESS

PRINCETON AND OXFORD

Published by Princeton University Press
41 William Street, Princeton, New Jersey 08540
99 Banbury Road, Oxford OX2 6JX

press.princeton.edu

All Rights Reserved

ISBN 978-0-691-23056-6
ISBN (e-book) 978-0-691-23058-0

British Library Cataloging-in-Publication Data is available

Editorial: Ingrid Gnerlich and Whitney Rauenhorst
Production Editorial: Jenny Wolkowicki
Jacket design: Chris Ferrante
Production: Jacqueline Poirier
Publicity: Maria Whelan and Kate Farquhar-Thomson
Copyeditor: Anita O'Brien

Jacket image: Early Bronze Age barbed and tanged flint arrowhead
(ca. 2500–1500 BC), courtesy of The Portable Antiquities Scheme

This book has been composed in Arno, Signifier, and The Future

Printed in the United States of America

10 9 8 7 6 5 4 3 2 1

To Gayle

CONTENTS

PREFACE

We live in an intelligence-driven culture. When we measure our mental abilities, we call the result an intelligence quotient. When we attempt to emulate our mental faculties by computer, we call it artificial intelligence. In searching the cosmos, we are seeking intelligent life. Some fear that machines that are more intelligent than humans would be an existential threat. This is all even though we are as far as ever from any widely accepted definition of what intelligence is. The evidence suggests that intelligence, much investigated as it has been, is an ill-conceived notion.

This volume defines an alternative: *educability*. Educability is the capability to learn and acquire belief systems from one's own experience and from others, and to apply these to new situations. A belief system here can be a scientific or political theory, a religion, a superstition, or a narrative in fiction. My subject is the extraordinary human facility to absorb and apply beliefs whether about science, creeds, or conspiracy theories.

The approach taken here in defining the capability of being educable is that of computation, or information processing. I have sought over several decades to understand, in the concrete terms of computation, what the mind does. What is the capability that humans have that has enabled us to understand and control the world as much as we do? The notion of educability emerged from these considerations.

The value of the computational viewpoint on human cognition is that it offers precise descriptions in terms of both *what* is being achieved and *how* it can be achieved. These more precise descriptions allow more complex notions, such as educability, to be formulated. Our current understanding of machine learning is an earlier instance of the success of such a viewpoint. It arose from a computational approach taken by this author and others to the human capability of learning.

In this volume I argue that the educability formulation fills widely recognized gaps in our understanding of human evolution, cognitive measurement, science-grounded education, human belief choices, and artificial intelligence. Viewing humans through this lens and encouraging further research along these lines is likely to lead to a better understanding of these issues.

I propose that educability provides the answer to the age-old question as to what distinguishes humans from other animals. More particularly, I am looking for the explanatory distinguisher, the quality that not only sets humans apart but also *explains* why and how we have been able to create the technological civilization that we have created. This *Civilization Enabler*, I suggest, is educability.

A correct characterization of this Civilization Enabler would set some actionable signposts for the future path of technology. Hence my characterization may also be interpreted as a proposal for machinery that can be educated. Some fear superintelligent machines. Perhaps we should be wary only of supereducated ones. While many argue that AI is some incomprehensible force, my treatment here provides some explanation of what it currently is and where it may go. The suggestion made is that educability is a good characterization of the core capability that AI might achieve.

I believe that the notion of educability as presented here has a new perspective to offer on several aspects of human affairs. In immediate human terms, educability encompasses the capacity to get educated. Since antiquity, many have advocated for the centrality of education to human life. Here this centrality is not only advocated but also explained. I present arguments showing that education is fundamental in a sense that was not so evident before and deserving of greater respect than society currently accords it.

Should the concept of intelligence in general discourse be retired in favor of educability? To date no one has found a definitive meaning of intelligence, while I am offering one here for educability. Should we be looking for educability in our political leaders?

I use the word "educability" throughout as a technical term for the computational capability that I am defining here. The term appears to be little used currently in any academic literature. In earlier decades it had been used as a criterion of a child's compatibility with standard school settings. My use of the term is different and more foundational. Here educability is an explicitly specified capability to acquire information and use it.

The novel content of this book is the *single computational model that defines the phenomenon that I am calling educability*. The model did not arise from experimentation but from theoretical considerations. The hypothesis I am offering is that this model will prove useful in understanding human phenomena relating to psychology, evolution, and education. The model being new, the hypothesis is, as yet, untested. I am not aware of existing empirical studies that bear decisively on it. I consciously avoid citing studies that are far from being decisive. In chapter 10 I make some suggestions for future studies that may shed light on how exactly the model relates to humans. I also refer to some findings in the psychology literature on questions that are explicit parameters of the model. However, this book is a theory-based work. For readers who are most conversant with empirical methodologies, this volume is an invitation to consider an alternative.

I believe that educability is a capability that all humans share. As with other capabilities, it can be expected that measures of it will vary among individuals, and these measures will be substantially influenced by the individual's past experiences. Research on methods of education that increase the general educability of a population would be worthwhile. Other than further research on educability for this purpose, the book makes no societal recommendations. For ethical reasons, this author would oppose any elitism based on measures of this or any other human capability. No currently controversial societal question is resolved here, and no unstated theory is implied.

In chapter 1 I introduce educability and its suggested role in enabling human civilization and, more generally, in human affairs. This chapter also introduces the computational approach. Chapter 2 discusses the evolutionary background, and chapter 3 the current relevance of identifying educability as the Civilization Enabler. Chapter 4 describes the foundation stones in terms of which the model will be defined. Chapter 5 explains the computational approach in more detail and seeks to justify its validity as a source of knowledge. Chapters 6 to 8 contain the explicit description of the educability model. In particular, chapters 6 and 7 describe Integrative Learning, a capability that many nonhuman animals also have, while chapter 8 discusses the additional step needed to achieve educability. Chapter 9 describes the rationale for the model. Chapters 10 to 13 discuss the many questions and perspectives that arise from the concept of educability. These perspectives include possible measures of educability, new directions for technology, some thoughts on education, and some dangers that humans face because of our unique abilities.

There is a brief summary of terminology at the end of the volume as a guide to the terms that are used in a technical sense. These terms can also be found in the index, where italicized page numbers point to where the terms are defined.

My goal is to present my own viewpoint on a phenomenon that I believe is central to human experience. My discussion will touch on issues within the scope of numerous academic fields, including human evolution, animal behavior, psychology, education, neuroscience, and computer science. The diverse existing literature that is relevant to this discussion would fill a large library. I can offer only regret for being unable in this brief volume to do justice to even the smallest part of that literature. Readers who believe that they have detected an omission are likely to be correct. On the other hand, these endless connections with other fields can be interpreted positively as a demonstration of the centrality of this concept of educability to human affairs.

THE IMPORTANCE OF BEING EDUCABLE

CHAPTER 1

The Civilization Enabler

Cleverness Is Not Enough

Once on a February day some chimpanzees in Belfast Zoo staged a spectacular escape. Their home was a high-walled enclosure opened by the primatologist Jane Goodall some decades earlier. The chimps had lived there uneventfully for many years in the presence of some hawthorn bushes. Shortly before that day, storms had weakened the branches of these bushes. This offered an unexpected opportunity. Soon after the keepers had left for the evening, it took only a few seconds for the daring escape to transpire. A broken branch was leaning against the high wall of the enclosure, reaching about a third of the way up. One of the chimpanzees, using the branch as a ladder, ran up along it, and with a running jump just managed to grab the top of the enclosure wall with one hand and pull itself up and over. Other chimpanzees followed. A sheer wall had been scaled.[1]

Such acts are striking examples of how bright animals can be. In any one case, it is difficult to tease apart the constituent capabilities that the animals are using to accomplish their feats. All the relevant history is rarely available. How did the broken branch come to be leaning against the wall? How much of the skills shown by the chimpanzees was innate rather than learned? Climbing along a branch and jumping to grab hold of something with one arm, as separate capabilities, must each have substantial innate components. Did one of the chimpanzees have previous experience with climbing a dead branch conveniently leaning against a wall? The surprise is not in either act alone, but in the combined event. In this instance, there was no evidence that any one of these chimpanzees had previously had serendipitous experiences that might have given it the idea. There is no reason to believe that any human intervention was involved. If one of the chimpanzees did indeed make a plan to do something never before witnessed by it, or by any of the other chimpanzees, then at least we have to raise our hats.

Crows are also famously clever. The abilities ascribed to them in Aesop's fable *The Crow and the Pitcher* are well recognized. In the fable, a crow drops pebbles into a jug of water, causing the water level to rise. Dropping in enough pebbles allows the crow to quench its thirst when the water level has risen sufficiently.

Researchers in animal behavior have explored this phenomenon systematically through experimentation. With food floating on the surface of water in a narrow-necked container, crows have been observed putting objects in the water. When the water level has risen sufficiently, the crows retrieve the food. When offered several such containers, they choose narrow containers over wide ones, narrow containers needing fewer pebbles to raise the water level. For the objects to add, crows will choose those that sink over those that float, reducing the volume of the objects needed. One does not need a discussion of what the crow really understands or intends to marvel at this behavior.[2]

We may wonder anxiously how we would measure up if presented with tests demanding similar cleverness. But here is the rub: however clever they may be, chimpanzees or crows cannot and will not develop a civilization like humans have. They will not construct a digital computer or travel to the Moon. Even if humans were not here as competitors, neither crows nor chimpanzees would be developing technological civilizations. So, what is the difference? What capabilities do humans possess that these clever apes and crows do not?[3]

I am seeking the capability that not only sets us apart from other living animals but also *provides an explanation* for why we have been able to create the technological civilization that we have. Identifying this civilization-enabling capability is my goal. I shall call this crucial human capability the *Civilization Enabler*.

The Civilization Enabler

It is believed that anatomically modern humans (hereafter referred to as *modern humans*) emerged in Africa more than three hundred thousand years ago, having evolved from within the broader genus *Homo* over the previous two million or so years.[4] The ancestors of modern humans already had significant capabilities that we do not see in other present-day species. By four hundred thousand years ago, these prehuman ancestors were hunting animals with stone-tipped spears, had controlled use of fire, and were cooking (see table 1). Stone tools go back much further, predating the *Homo* genus altogether.

Despite these impressive capabilities already available at the beginning, cultural progress in the three-hundred-thousand-year history of modern humans appears to have been slow initially. Eventually, by around fifty thousand years ago, marked changes had occurred, sometimes called the "cognitive revolution." Art and more complex tools had appeared. Our transformative physical impact on the planet in the form of large settlements or buildings took even longer. These made their mark only a little more than ten thousand years ago. Most of what we would recognize as civilization is even more recent.

Progress toward modern civilization was therefore glacially slow at first. What prompted the more recent advances, and why were they only manifest so late in the history of our species? What caused this delay of three hundred thousand years, or more than ten thousand generations, before our civilization-enabling genes kicked in? Why did they kick in so decisively when they did?

It is possible that this slow progress tracked genetic changes that were happening in humans over these past three hundred thousand years. However, the search for genetic changes that closely correlate with advances in human activity has not yielded anything conclusive. One systematic search sought to find a genetic mutation that spread through the population during this period and that all humans now share.[5] Such an occurrence would be called a *complete selective sweep*. Given the genomes of a set of humans, one can apply statistical methods to estimate for each segment of the genome when the most recent common ancestor of all that genetic material lived. This study estimated that the most recent common ancestors of all the segments searched lived more than three hundred thousand years ago. This suggests that there was no genetic mutation that swept through the whole of the human population in that period. The authors note that their study would have missed mutations occurring in parts of the genome not searched and epigenetic mutations.[6]

It is not that the intervening period was devoid of known genetic events. Humans migrated between continents, and there was some interbreeding with related species, the Neanderthals, the Denisovans, and perhaps others. There have been specific physiological changes in large populations, such as increased tolerance of cow's milk and recent changes in height. Tolerance of cow's milk is due to genetic mutations that spread through only a part of the human population. This is not what we would be looking for as far as the Civilization Enabler that all humans share. Recent changes in height are believed to be *polygenic adaptations*, in which the relative proportions of a set of previously existing gene variants change, without new mutations occurring. If there have been genetic changes in the past three hundred thousand years that have affected our cognitive abilities, they may have been of this polygenic kind where the proportions of different gene variants change in the population, but no novel mutations occur.

It is therefore possible that the critical civilization-enabling mutations appeared with or even much before the emergence of modern humans, but their consequences have played out more slowly. The genetic mutations that increased tolerance for cow's milk gave rise to the selective advantage of a broadened diet that could be enjoyed immediately. In contrast, the advantages of the capability that enables civilization, at least at the most spectacular levels, may have taken thousands of generations to fully manifest themselves.

Now I come to the main proposal of this book. It is a new hypothesis, one that is respectful of these last observations. The hypothesis is that the sought-after Civilization Enabler is what I shall call *educability*. This is a new notion that I shall define more precisely in later chapters. Most basically, educability is concerned with our abilities to *acquire* beliefs, and to *apply* them in specific situations, where the acquisition process allows both the acquisition of beliefs *explicitly described* to us by others and an ability to learn beliefs by *generalizing* from our individual experience.

Table 1. A selection of events in the cultural evolution of our genus *Homo* for which supportive physical evidence has been claimed

2,600,000	Stone tools (Gona, Ethiopia) Possibly a predecessor of *Homo*
1,700,000	Stone hand axes (Tanzania) *Homo erectus?*
1,500,000	Control of fire (Kenya) *Homo erectus*
500,000	Abstract markings: a zigzag engraving on shell (Indonesia) *Homo erectus*
500,000	Stone-tipped spears (South Africa) *Homo heidelbergensis*
320,000	Long-distance transport of obsidian for fine blades and points, and ochre for pigments (Kenya)
200,000	Adhesive: birch tar for hafting stone tools (Campitello, Italy) Neanderthal
170,000	Widespread use of clothing (Africa)
142,000	Symbolic ornaments: marine-shell beads (Morocco)
120,000	Burial of dead (Qafzeh Cave, Israel)
100,000	Mixing and storing pigments (Blombos Cave, South Africa)
90,000	Bone harpoons (Semliki River, DR Congo)
75,000	Jewelry fashions: shifts in styles of threaded shell beads (Blombos Cave, South Africa)
64,800	Symbolic cave paintings (La Pasiega Cave, Spain) Neanderthal
50,000	Eyed needle, made from bone (Denisova Cave, Siberia) Denisovan
48,000	Self-medication with natural pain-killer and antibiotic (El Sidrón, Spain) Neanderthal
45,500	Representational art, a red-ochre composition of warty pigs (Leang Tedongnge, Sulawesi)
42,000	Musical instruments: bone and ivory flutes (Swabian Jura, Germany)
42,000	Fish-hooks, manufactured from broken shell (East Timor)
40,000	Figurative sculpture, an ivory figurine with lion's head and human torso (Hohlenstein, Germany)
35,000	Fully human sculpture: a mammoth-ivory "Venus" figurine (Hohle Fels, Germany)
32,600	Food-plant processing, of dried wild oats with grindstones (Grotta Paglicci, Italy)
30,000	Woven fabrics, made from dyed fibres of wild flax (Georgia)
24,000	Poison arrows, with wooden ricin applicator (Lebombo mountains, South Africa)
23,000	Fisher-hunter-gatherer brush huts (Sea of Galilee, Israel)
23,000	Domestication: dogs from gray wolves (Siberia)
20,000	Pottery vessels (Xianrendong Cave, China)
15,000	String instrument: the musical bow (cave painting at Trois Frères, France)
14,400	Baking bread: unleavened flatbread from wild einkorn and club-rush tubers (Shubayqa, Jordan)
11,600	Monumental ritual art (Shigir, Siberia): 5-meter-tall plank carved with human forms and signs
11,500	Cultivation of wild barley and oats around village settlements
11,500	Monumental temple (Göbekli Tepe, Anatolia)
11,000	Continuous settlements (southern Levant)
8,500	Mining of metal, to heat, hammer, and grind into tools, projectile points (Great Lakes, North America)
6,000	Earliest board games (Egypt)
5,500	Domestication of horses (Central Asian steppes)
5,400	Wheeled wagons (Germany, Slovenia, Near East)
5,300	Numeral systems: pictograms of economic units (Uruk, Mesopotamia)
5,200	Full writing (cuneiform in Mesopotamia, hieroglyphics in Egypt)
4,650	Massive stone monuments (Egypt); contemporaneous pyramids (Peru) and megalith (UK)

Abstracted from a table constructed by C. Patrick Doncaster, "Timeline of the Human Condition—Milestones in Evolution and History," https://www.southampton.ac.uk/~cpd/history.html, which also gives sources. Used with permission. *Note:* Estimated dates given are years before the present. Future findings may indicate earlier dates.

The sets of beliefs that humans can acquire I call *belief systems*. Some use the phrase "belief system" in the narrower senses, such as religious beliefs. Religious belief systems are examples of what I mean by this phrase. They are certainly among the more ancient belief systems of which we have evidence. The fact that there are many major religions and even more variants of each, I regard as evidence of the human facility for complex sets of abstract beliefs. There is no evidence that nonhuman animals have anything that corresponds. But my use of the term "belief system" here will also encompass other systems of beliefs, including myths, stories, methods of doing things, and the sciences. The commonality among them that is exploited here is the commonality in the ways we mentally acquire and apply them, and not in the degree to which they are, or we believe them to be, true.

Some groups of humans have been genetically isolated from the rest for a long time. It has been argued that the San from southwestern Africa have been largely isolated for more than one hundred thousand years.[7] However, San culture shares with humans across the globe all the basic features that I shall associate with educability. Like other groups, they have origin myths, deities, and stories about their deities, which all involve complex belief systems that individuals acquire from others. They also have sophisticated knowledge that enables them to survive in their desert environment. For example, they traditionally hunt game using arrows poisoned by extracts from the roots of certain not so easy to find desert plants. This hunting technique must have been difficult for an individual to invent, given that it uses knowledge that would be challenging to learn from experience. Once invented, however, it is easy to pass the technique down from generation to generation.

In acquiring their culture, the San therefore needed all the capabilities that are essential to educability, namely, the ability to learn from experience, combined with the abilities to acquire complex theories from others and to apply knowledge gained by either method to new situations. The San's traditional culture fully satisfies the requirements of our definition of educability. If the San have been totally isolated for one hundred thousand years, they could not have acquired mutations occurring elsewhere in the human population during that period, suggesting again that all the genetic requirements for human civilization were already in place for quite some time in the history of humans. As always, some caution is needed as there are uncertainties about how to interpret the evolutionary evidence. Proving total isolation for any group would be difficult, and, indeed, there is evidence that the San were not totally isolated genetically during this period.[8]

Fortunately, to make the case for educability, I shall not need to make *any* assumptions about the details of evolutionary history. The educability hypothesis is consistent with the idea that the development of human cognitive behavior within the past three hundred thousand years largely followed its own slow course, set on its way by our species genetically having the educability capability from the beginning. The hypothesis is minimalist in not assuming that any specific changes occurred in the genome over the course of human history.

One can make several speculative arguments for why educability may be exceptional among human capabilities in the slowness of the pace of its impact after it first emerged. A nongenetic argument is that at the beginning, when little knowledge had been accumulated, the benefit of communicating it to others was small. A genetic argument is that this capability may be the result of polygenic adaptation and dependent on what fraction of a large set of genes favor it. In that case, the fraction of relevant gene variants that favor educability may have been initially low in the population. The benefit of receiving knowledge from others is small if others have low educability and therefore little knowledge to share. For both these reasons, the selective advantage offered would be small initially. It is therefore plausible that following the emergence of the capability for educability, there would be a lengthy period in which the average educability and shared knowledge of the population both increase only slowly. Like snowballs that start rolling, their initial movement is almost imperceptible.

For many years evolutionary biologists and others have been seeking to track down in detail the individual developments that occurred in the emergence of our civilization. They use DNA evidence, the evidence of artifacts from the past, and knowledge of recent human culture. My emphasis here is different. I focus on the *current result* of human evolution and aim to characterize one aspect, which I call educability. If we are to understand our defining evolutionary story, it would seem hard to evade the question of *what* is the defining cognitive capability that we have and that has evolved.

Knowledge Accumulation

The power of educability derives from the fact that the knowledge an individual can acquire if transference from others is possible is incomparably greater than what one could have discovered from one's own experience and efforts alone. Educability offers the individual the enormous power of having knowledge that took multitudes to discover over many generations. This power is not available to species that lack educability.

Language, speech, and the practice of recording information on tablets or paper accompanied the development of civilization and clearly facilitated it. But what made these abilities and technologies useful? My answer is that it is our educability. Technologies for recording knowledge and abilities to communicate have limited power in and of themselves. To individuals who are educable, however, they offer the power of using the knowledge on an unlimited scale.

In the sixteenth century Tycho Brahe observed the sky at night and collected systematic data on the positions of the planets over several decades. For this, Brahe used an observatory that is thought to have consumed for a time a sizable portion of the government expenditure of his country, Denmark. After Brahe's death, Johannes Kepler, his former assistant, used this data to deduce that the planetary orbits were elliptical. To start, Kepler needed only to read Brahe's tabulated data, without having to repeat the observations or expenditures. Kepler also had to do much more. Essential to his discovery were mathematical notions that he learned from others. He needed the notion of an ellipse, which was known almost two thousand years earlier in Greece. For his third and last law, Kepler needed to do complex calculations. For these he used the method of logarithms, about which John Napier had recently published in Scotland.

Kepler was able to exploit all this knowledge, both the data and the mathematical principles, which he had no chance of observing and deriving from scratch by himself. Through a combination of formal education and self-education, he put himself in the position of being able to use what others before him had obtained with lifetimes of effort. The understanding that planetary orbits were elliptical remains one of the crowning achievements of humanity, one that has had decisive impact on the subsequent development of science. Kepler's facility in absorbing and applying previously obtained knowledge made it possible.

To the question of what genetic changes happened between the first emergence of modern humans more than three hundred thousand years ago and the visible large-scale products of civilization around ten thousand years ago, our answer is that nothing dramatic *needed* to have happened. It may have just taken a long time for educability, which the earliest humans could have possessed, to manifest its full power. For hunter-gatherers without much physical security or much spare time to teach and learn, the opportunities for flaunting the power of educability might have been slight.

By early seventeenth-century Europe, the environment was quite different. An individual like Kepler could exploit the information he had learned from others with enormous consequence. In the present day, information is disseminated around the world at an ever more feverish rate, and scientific discoveries are being made at a correspondingly ever more rapid pace. This pace is maintained with the help of universal education and digital technology. To exploit educability to the full, both of these may be essential.

It had taken a long time before the cumulative value of the knowledge gained across the planet was great enough to become self-sustaining as a process and spawn the technological civilization we have today. Through a gargantuan multigenerational effort, the Civilization Enabler has given us a good understanding of the physical world, and a capability to transform it.

A Computational Approach

What exactly is the nature of this Civilization Enabler? Its fruits are easy enough to see: culture, the arts, knowledge, science. Here I am interested in going further. I want to understand the human *capability* that gives rise to these fruits.

The notion of educability as I define it did not arise from and cannot be defined in terms of physics, chemistry, biology, or the social sciences. Nor can it be justified by the methodologies used in these sciences. This is hardly surprising since its subject matter is information processing, which is not the focus of these sciences.

While the twentieth century saw unparalleled developments in the classical sciences, equally important and particularly in the work of Alan Turing in the 1930s, it saw the birth of the science of information processing. By that time, it was commonplace not to marvel that physical concepts that are not visible, such as energy or electric charge, could have useful meaning. The fact that the same held for notions of information processing and computation, terms that I shall use synonymously, was startling news.

The import of this news was well understood by the early pioneers of computing, namely, Turing himself and John von Neumann. They sought immediately to use computation to study biological phenomena, such as the brain, cognition, and genetics. Each of these phenomena involves the transformation of information. Focusing on the information processing rather than the physical realization became a viable and necessary approach toward understanding these once a scientific approach to information processing had come into view.

I, and many others, consider it self-evident that if we are to understand how the brain works, we will need to understand it in terms of information processing. Some skeptics have suggested that information processing is only a metaphor. They point out that there is historical precedent for comparing the brain with the most complicated and prized machines of earlier eras, such as the camera. Perhaps once again we are just comparing the brain with the most complex machine that we happen to have, the computer. Perhaps computation is just one of many possible metaphors for the brain and there is nothing much to choose among them.

I would say that this view is mistaken. Physics is more than a metaphor for the physical world. It relates to falling apples and the motion of the planets more than as a mere metaphorical narrative. It seeks to *explain* falling apples and the motion of the planets. Whenever a gap appears between the description offered by physics and the observed behavior of falling apples or moving planets, every effort is made to update the physics. Physics is more than a metaphor because its *ambition* is more than that. It is always willing to change and improve so as to be more useful in explaining reality. It has usually succeeded.

The case is the same for computer science as it relates to the world of information processing. Long before humans, life on Earth was processing information. The copying and mutations of the DNA that occurred between successive generations of bacteria, animals, and plants is information processing. With the evolution of nervous systems, information processing came to be carried out in brains on a yet different and even more massive scale.

Computer science is more than just a metaphor for the world of information processing. Its ambition is all-encompassing in aiming to explain *every* kind of information processing that is possible, whether in biology, silicon, or some other realization. As long as gaps are found between the description that is offered by computer science and a real-world information processing phenomenon, efforts will be made to update computer science in order to resolve the gap. It is willing and able to change and improve.

Why Computation?

In a nutshell, computation is used here to provide *concrete* descriptions of processes, such as of learning and education. There are two senses in which I will need this concreteness. The first sense is that of *precisely specifying the outcome* of a process—defining, for example, exactly what the outcome of a learning process needs to be if we are to declare that learning has been successfully achieved. The second sense is that there is a precise *step-by-step description*, possibly as a computer program, of how this outcome *can* be achieved using reasonable resources—one cannot ascribe to the brain or other nature phenomena capabilities that one does not know how to realize concretely by any means in this universe.

My intention is to define a notion that is good for more than a coffee table discussion. I am claiming that educability is a useful scientific concept. To make that case, I will need to specify the nature of educability quite precisely.

I regard both educability and education as phenomena of computation. In the course of education, information is presented, whether as the description of a specific situation or as an explicit description of a general belief. The result of the presentation will be to make a difference in the student's subsequent behavior as compared with the past. The change in behavior will be attributable to a change of the *state* of the student, realized as some physical change in the brain that persists for some time.[9]

Computation is about changes of state that can be realized by step-by-step processes. Physical systems also change state—if you boil water there is a change of state. In a computer or a brain, there is extreme flexibility in the realizable state changes and in their possible effects. In both computers and brains, there are billions of parts that at any instant have some state. Each state arises as the cumulative effect of past experiences. Each state can influence future behaviors. In computers certainly, the state of a single one of the billions of elements, namely, whether it is a 0 or a 1, can have a decisive effect on later behavior. The way in which the effect depends on the state can be arbitrarily complex.

Computation encompasses all ways experience can cause changes of state and, in turn, state can influence later behavior. It is this expressiveness and flexibility of computation that is so useful for describing cognitive capabilities.

The idea that many of the unresolved secrets of biological phenomena, such as cognition, lie in the world of computation needs some discussion. I will try to break this down further in chapter 5. Progress in unraveling these secrets has been slow, and the difficulty of treading this path has often been underestimated. In a talk at a conference banquet in 1957, the artificial intelligence pioneer Herbert Simon made four bold predictions.[10] The first was that "within ten years a digital computer will be the world's chess champion, unless the rules bar it from competition." As it happened, it took forty years for computer chess to approach the needed level, and the International Chess Federation rules have been barring its participation.

Simon's fourth prediction is the most relevant to the discussion here. He predicted that "within ten years most theories in psychology will take the form of computer programs, or of qualitative statements about the characteristics of computer programs." Well, even an extra half century has not been enough for this to happen. I believe the reason is that the necessary connection between psychology and computation is more subtle than Simon's statement suggested. This book is about some of the nuanced connections between the two. Explaining these connections, and why they work, is one of the tasks I take on here, and one I will keep coming back to.

CHAPTER 2

Where We Come From

Darwin's Challenge

Darwin issued a challenge, and with great clarity, in his *Descent of Man* (1871), where he wrote that "there is no fundamental difference between man and the higher mammals in their mental faculties." He then proceeded to give extensive arguments in support of this proposition. He explained that "the difference in mind between man and the higher animals, great as it is, is certainly one of degree and not of kind." He pinpointed the difficulties in any such discussion as "Self-consciousness, Individuality, Abstraction, General Ideas, &c.—It would be useless to attempt discussing these high faculties, which, according to several recent writers, make the sole and complete distinction between man and the brutes, for hardly two authors agree in their definitions."[1]

The scientific community has not accepted Darwin's position on this as the last word and has made great efforts in the century and a half since to find a specific explanation for the apparent uniqueness of human behavior. As already mentioned, many such efforts attempt to use Darwin's theory of evolution. These approaches seek to understand how evolutionary pressures worked on the biological, social, cultural, and ecological conditions in which our ancestors found themselves, to bring about successive changes. This approach is sometimes referred to as *gene-culture coevolution*.[2] The idea is that cultural innovations, such as novel stone tools, would give a selective advantage to those best able to make those new tools, and, in the reverse direction, that any such resulting genetic changes would in turn have the effect on culture of enabling the making of even better tools. The aim is to understand the history of the genus *Homo* over the past several million years from the viewpoint of this interplay between culture and genes.

Certainly, in the last several million years our ancestors underwent many changes in both biology and culture: we started to walk on two feet, our brains grew greatly in size, and we acquired language. However, the evolutionary explanations for these changes remain elusive, if only because of the lack of direct evidence. While larger brains may offer advantages, it is less clear how selection rewarded this massive biological investment in the short term. The benefit must have been more than just the ability to find sweeter fruit. One theory is that our ancestors had complex social relationships, and those with larger brains could manage these better.

Educability offers a computational approach to Darwin's question as to "the fundamental difference between man and the higher mammals in their mental faculties." It spells out quite explicitly the hypothesized defining characteristic of our present-day mental faculties and in doing so aims to explain what evolution has wrought. As mentioned earlier, it would seem useful to understand the *end-product* of evolution before we ask about the *how* and *when* of its production.

The Brain

Any cognitive capability we attribute to humans will ultimately need an explanation in terms of brain function. Unfortunately, while we know much about the brain's biochemistry, we still know little about how it processes information. We do not know, even to orders of magnitude, how many neurons are involved in our memory of the breakfast we had this morning. Hence any detailed discussion of how the brain might support the capabilities discussed here is for now only at the level of plausibility.

The general structure of our brain is similar to that of other primates. The idea that there is some dramatic difference between the human brain and that of the other apes, such as the presence of a special brain structure, was circulating at the time that Darwin was developing his theory of evolution. For example, Richard Owen, the originator of the word "dinosaur" and founder of the London Natural History Museum, argued strongly for this idea in the 1850s. To date, no evidence for any differences in organization between the brain of humans and apes that are glaring enough to have been detectable in the nineteenth century have been found.

On the other hand, there has been ample time since our ancestors genetically diverged five to seven million years ago from the ancestors of other living apes. Clearly, much could have happened in that period. One clear difference in the brain of humans as compared with apes is quantitative rather than organizational: the human cortex is three times larger than that of other living apes. This may not be the only crucial quantitative difference. Other candidates are the number of connections per neuron and the synaptic strengths of the connections, which quantify the amount of influence a neuron has on its neighbors.

Each of these quantities has influence on the computational power of systems of neurons.[3]

There are also differences in organization that are subtle but may be significant. The prefrontal cortex, which is associated with decision-making and short-term memory, is proportionately a larger fraction of the brain in humans than in other apes.[4] The different brain areas are richly connected to each other in all primates. Exactly which areas have direct connections with which others differ in humans as compared to our relatives.[5] Our brains are more asymmetrical in function than those of other primates, the two hemispheres being more different in what they do. Such functional asymmetries exist, for example, in speech production. Some asymmetries in the physical connection patterns between the two hemispheres in humans have also been noted.[6]

Changes in the brain are known to have occurred in our genus *Homo*, though we do not know their effect on our mental capabilities. Some changes during the last two million years are preserved in the fossil record. Endocasts—replicas of the inner surface of the skull—reveal certain surface features of the brain. Such features include surface grooves, known as *sulci*. From fossil skulls one can also determine the location of *sutures*, where the main parts of the skull are held together. It turns out that one such sulcus, the *precentral sulcus*, crosses one such suture, the *coronal suture*, in the other great apes, but not in modern humans. A recent study examined fossil skulls to see at what stage in human evolution this change happened.[7] It was found that certain 1.8-million-year-old fossils of the genus *Homo* do not show this pattern, but by three hundred thousand years ago the modern pattern had emerged in modern humans.

The evolutionary history of the genus *Homo* is known to have been highly convoluted, with several species emerging and vanishing. Our knowledge of this is incomplete because the fossil record is incomplete. This study of the *precentral sulcus–coronal suture* crossover does show that at least one brain development was not complete when *Homo* first emerged and occurred during the subsequent history of our genus. Changes that occurred within the most recent three hundred thousand years of the existence of modern humans are also known. In particular, the shape of the brain has become more globular.[8]

We cannot expect that the anatomy of the brain will by itself reveal the Civilization Enabler. As a first step, we need a specification of what the Civilization Enabler does.

The Difficulty of Invention

The creation of a technologically advanced civilization such as ours does appear to take some doing. There is no evidence that any of the millions of species that have existed on Earth, other than those in the genus *Homo*, had or have the capability to produce anything like it.

Even for our species, our technology has taken an almost inexplicably long time to develop. It needed long sequences of discoveries and inventions, and each step seems to have taken an inordinately long time by the standards of change in our current world. It is difficult to evaluate the level of difficulty of any one discovery or invention since the magnitude of the hurdle that was overcome relates to what exactly was known before and what questions people had reason to ask.

For example, was the invention of the wheel for transportation a difficult task? As far as we currently know, this invention came quite recently, less than six thousand years ago. Early near-contemporaneous evidence of wheels for transportation has been found at multiple sites in Europe and the Middle East. Although wheels on vehicles are now everywhere around us, imagining one for the first time was no small feat. Circles and circular objects may have been ubiquitous, as the Moon or in flowers or as grinding stones. Wheeled toys had already existed. Wheels for transportation are different. First, the invention would require a need. There has been much discussion of why wheeled vehicles had not been known in the Americas before the arrival of Europeans. The rough mountainous terrain of the Andes, for example, may not have offered a sufficient incentive for going to the trouble of crafting a wheeled vehicle. Second, a mechanism, such as a shaft fixed to a vehicle and a wheel freely rotating around it, had no precedent and needed to be imagined. Third, a technology to make it was needed. Wood is a suitable material and was widely available. Some have suggested that metal tools such as chisels had to be invented first.

Our civilization encompasses thousands of such inventions. Every day we use many of them without thinking. The invention of each one was the fruit of both labor and ingenuity. Each built on earlier inventions and depended on efficient communication of knowledge. Currently news of good inventions spreads around the world like wildfire.

Distinguishers, but Not Civilization Enablers

Historically, the quest to determine what is unique to humans took on a new meaning after Darwin. Long lists of proposed differences between humans and other animals were compiled, some containing hundreds of items. One current list contains more than six hundred different proposals, each of which has an academic literature regarding the possible differences between humans and the other great apes.[9]

For each candidate for a human distinguisher, we can discuss whether the feature or capability *is* unique to humans. Tool use was at one time thought to be unique to humans but is now known to be shared also by other primates and some birds.[10] New Caledonian crows are particularly famous for using twigs as tools. They are among a very few nonhuman species that have been observed crafting tools. They will find a suitable branch, remove unwanted leaves, and with their beaks trim the shaft and finally sculpt a terminal hook. A stick with such a hook is particularly effective for getting at insects hidden in deadwood and vegetation.[11]

Making comparisons between humans and other animals needs caution. Rutz and St. Clair have suggested that tool use for these crows may have evolved from ancestors who, like woodpeckers, retrieved their food with their beaks without any tool.[12] Experiments on captive-bred New Caledonian crows reveal a strong genetic predisposition to tool use. The authors argue that tool use here may be a specialized evolutionary development, analogous to the development of a specially shaped physical beak, rather than a manifestation of more general tool-using abilities. In other words, tool use for these crows was acquired via evolution like the use of a particular tool, such as a chisel, might be learned by a human during life. The phenomenon of tool use in New Caledonian crows may therefore have little commonality with that in humans, who are born with a much broader and flexible potential to learn the use of diverse kinds of tools.

On the other hand, there are also reports of animals using tools where apparently the individuals needed to discover how to use the tools by themselves. A parrot in New Zealand with a missing upper beak reportedly uses a stone held between its tongue and lower beak to preen itself.[13]

The question of whether a capability is unique to humans is therefore already quite complex. Fortunately, for the argument here, our focus is on the other issue: Even if it is unique to humans, does the capability *explain* how it makes civilization possible? Some have proposed that humans alone possess eyebrows. Whether or not eyebrows are unique to humans, there is no argument known for the proposition that the possession of eyebrows is critical in the development of civilization. The same applies to most of the other items found in the conventional lists of proposed distinguishers. For example, bipedal posture, brain asymmetry, cooking, complex tools, domestication of animals and plants, all appear to have accompanied the emergence of our ancestors, but none of them separately, or even together, explains why we are capable of creating advanced cultures and civilizations. While domestication of animals and plants may have played a key role in making possible the large communities and cities that now pervade our planet, there is no argument to explain that if you have domestication then that will lead to figuring out how to go to the Moon.

A few of the proposed distinguishers are likely to be *related* to the Civilization Enabler. Large brain size relative to body size is likely to be relevant, simply because, through evolution from our common ancestors, that is the most obvious biological feature in which we have come to differ from the other living apes, and we know that the brain is the center of our cognitive processes. The crows and parrots we so admire for their cleverness also have large brains.[14] But in the end, we also want to know how exactly this larger brain is used to enable a complex civilization.

Other proposed human distinguishers have been language, reasoning, generalization, social cognition, culture, and consciousness. No consensus has arisen about any one of these, and some other investigators have also concluded that none of these commonly discussed traits accounts for human cognitive uniqueness.[15] Discussion of behaviors that humans have in abundance is, of course, relevant here and provides the subject of many noteworthy books. Richerson and Boyd, as well as Laland, emphasize the role of culture and cultural learning in human evolution.[16] Henrich also takes an evolutionary approach and emphasizes social learning and collaboration as being characteristics of humans.[17] Tomasello takes a developmental rather than evolutionary view and focuses on cooperative thinking and shared intentionality.[18]

The theory offered here does not contradict the particulars of these other approaches but takes a different view of what is explanatory of civilizations. Educability is offered as the basic human cognitive capability. The implied connection with these other approaches is the suggestion that this capacity would be supporting the more particular behaviors and phenomena that are shown in abundance by humans and not by other living species.

The educability proposal also suggests a simple approach to the how and when of human evolution that is alternative to gene-culture coevolution. Suppose that our defining cognitive characteristic is such that having more of it always gave one a (possibly weak) competitive edge over those who had less. Then a steady evolutionary pressure toward having more of it needs no further explanation. The many stages of the tortuous history of our ancestors may each have had limited long-term impact on the overall course of evolution, as compared with the steady selective pressure toward enhancing this one characteristic. I claim that this argument makes sense when the characteristic, educability, has a definition.

The aim of my discussion of evolutionary history here is to show that the educability concept is consistent with a wide range of possibilities as to when humans first acquired their various capabilities. The gene variants needed for the Civilization Enabler may have been in place two million years ago. At the other extreme, it is possible that there were some critical, but yet to be identified, genetic or epigenetic changes in the past three hundred thousand years. Presumably, the Civilization Enabler had some selective advantages to offer as soon as it emerged. It seems that it had, as if spring-loaded, unimaginably more in store.

CHAPTER 3

Where We Have Got To

The Civilization Enabler did not vanish when civilization had been enabled. For better or worse, it remains with us, decisively controlling our lives. This chapter discusses its current relevance. First, I discuss the rise of educational activity in the past few centuries. This I take as a clue to the centrality of our ability to be educated. Despite the enormous resources expended on it, however, education as a process is not well understood. Currently it is conducted more on the basis of best practices than on accepted scientific principles.

I go on to discuss how education has effects that are beyond individuals. Communities of educated individuals produce more than the sum of their parts. Finally, I discuss why the conventional notion of "intelligence," which has been playing a pervasive role in education, should not be identified with the Civilization Enabler or our uniqueness.

Pervasiveness of Education

Many societies have put significant effort into educating their young. While the teaching of practical skills has always had a role, some have defined education much more broadly than that. G. K. Chesterton wrote: "Education is simply the soul of a society as it passes from one generation to another."[1]

There has been much debate about the purpose and benefits of education, as well as the methods for delivering it. For Plato, education was central to an ideal society. Plato addressed the question of how best to organize society through the power of education. In his *Republic*, he proposes elite leaders, the philosopher guardians, who would undergo a rigorous process of education not completed before the age of fifty. Many societies have used education to enforce a common culture or religion. Using education to help create new knowledge at a steady pace was not among explicit goals until recently, though some societies did excel in producing new knowledge.

In the diverse world that we inhabit today, it is perhaps surprising that there is any one thing on which there is near unanimity: that one thing is universal or compulsory education. Almost every country ensures that all children are educated, usually for many years. Compulsory schooling may start as early as age three and end as late as age eighteen. The nature of the schooling and the curriculum vary widely, but this near unanimity is still remarkable given the enormous commitment of time and resources that universal education requires of a society.

Universal education has several roots. Almost three thousand years ago, Sparta instituted the *agoge*, which provided all sons of citizens rigorous training in the art of war. In Judea, in the first century, according to the Talmud, the high priest Joshua ben Gamla ordained that "teachers of young children should be appointed in each district and each town, and that children should enter school at the age of six or seven."

Universal education for both boys and girls existed in the Aztec Triple Alliance, which flourished between 1421 and 1528, centered on the city of Tenochtitlan, in present-day Mexico City. Education until the age of fourteen was the responsibility of parents, but under the supervision of the authorities. At age fifteen, education became compulsory.

In North America, Massachusetts was the first British colony to mandate schooling. The Massachusetts General School Law of 1647 ordered that every town with fifty households appoint someone to teach children to read and write, with wages paid by the parents or town inhabitants. Every town of a hundred households was required to set up a grammar school with a teacher who could instruct the youth "so far as they may be fitted for the university." Towns that failed to comply had to pay a fine of five pounds, which several did. The preamble to this so-called Old Deluder Satan Law starts: "It being one chief project of that old deluder, Satan, to keep men from the knowledge of the Scriptures." While the law had a specific religious motivation, the legislators provided children with the broader capabilities to read and write.

In Europe compulsory education started in Prussia and Austria in the eighteenth century. In 1763 Frederick the Great of Prussia issued a decree that required that all young citizens be educated by mainly publicly funded schools from the age of five for about eight years. In Austria in 1775, Empress Maria Theresa mandated that all children had to attend school between the ages of six and twelve.

Character is sometimes identified as a goal of education—and rightly so—even if this goal is hard to define. There is little consensus on how formal education can be successful in this area. It is difficult to sustain claims of transformational success for any one known approach. Education may well influence character, but the direction of influence may not be in line with intentions. Tom Nichols, author of *The Death of Expertise*, has proposed that universal education, despite all its beneficial effects, has had unintended consequences. He claims that it has produced a population that overrates its abilities to analyze complex problems and in consequence undervalues the judgment of experts, such as medical doctors. He suggests that our education system produces narcissistic individuals, who dismiss the judgment of experts because they think that they know better.[2] If current educational practice has had this unintended side effect, then we do have a problem. Character has to do with an individual's belief system regarding their relationships with others. How the belief system that manifests itself in narcissism comes into being is worthy of research. Inborn characteristics, learning from experience, and ideas acquired through instruction all have a role. How these combine together to create a narcissistic individual is an important question and within the broader scope of study suggested by the educability framework.

With the rise of universal education there has been reason to ask: What should be the principal goal of education? Aristotle had expressed a firm opinion much earlier: "No one will doubt that the legislator should direct his attention above all to the education of youth. . . . The citizen should be molded to suit the form of government under which he lives."[3] There have been proposals for many alternative goals besides the molding of citizens. Is the goal to develop character? Is it to maximize the individual's contribution to society? Is it to develop critical thinking? Is it to advance knowledge? Is it to produce leaders?

The answer—an amazing one—is that any of these, and many more, can be entertained. The infinite possibilities demonstrate the astonishing and unique power of this ability to be educated. *What then is this phenomenon with such limitless possibilities?* This is the question that the study of educability can illuminate.

Education in the Large

What is education beyond the transference of knowledge between individuals? Is it no more than like photocopying an encyclopedia? Instinctively we know that there must be much more to it than that. A starting point is the observation that, as individuals, we cannot do everything for ourselves. We each rely on generations of others to have produced the knowledge that will form the basis of *our* knowledge. We each need a decade, and sometimes more, of education to attain a substantial level of familiarity with existing knowledge. We are quite helpless as isolated individuals. We should therefore not be dismissive altogether of the encyclopedia photocopying metaphor. The effect of education on a society, however, is much more than this metaphor suggests.

The Apollo program that landed a human on the Moon is worth reflecting on. At one level, it was a triumph of organizing into a single project some four hundred thousand people with varied expertise. Individuals brought to the task knowledge that they had mostly acquired from others and could not have discovered for themselves. Equally striking, the group had knowledge that was too great for any one member to have acquired. This was a spectacular example of how humans have succeeded in exploiting education not just at the scale of an individual but also on the scale of a large, diverse group.

In fact, the organization of these four hundred thousand people for the common purpose of getting a person to the Moon is a suitable metaphor for the education of a single person. The enormous organizational effort expended in the Apollo program is analogous to the organizational effort made over several centuries toward creating the education system and teaching materials for a single present-day student studying for an advanced science degree.

While the amount of knowledge that can be taught to a single person, or to four hundred thousand people, is impressive, this encyclopedia metaphor does not capture all of education. The educability notion extends it in two specific ways. In one direction, educability requires that the individuals who have this knowledge can also *apply it* to new situations. Applying knowledge may involve the complications of combining several pieces of knowledge appropriately, or, in other words, doing some reasoning on them. In a second direction, the notion of educability incorporates the ability to acquire beliefs based on one's own immediate experience of the world. This is an essential part of human cognition. A piece of knowledge that we keep transferring to one another must have had its origin in one individual first conceiving it.

The success of education in advancing science and technology has varied widely in different societies. It may be that education systems do not all serve scientific progress equally. We do not understand well how differences in education systems affected other differences among societies. Given the time and resources societies put into education, perhaps it is time that we did.

Educability Is Different from IQ

A story, possibly apocryphal, about a computer company in the 1970s goes as follows. For hiring each cohort of university graduates, the company administered a battery of psychological tests over a day and a half. After some years, the company investigated which of the administered tests correlated best with the perceived work performance of those hired. They found that the test that was most predictive was the one where a list of 0s and 1s had to be copied accurately from one column to the next.

The company was not hiring graduates for copying 0s and 1s. Nonetheless, this test appears to have served as a proxy for the qualities it was looking for. One can ask the same question about IQ tests, which have existed now for over a century. No one would claim that the questions on these tests are the ones that employers need solving. IQ tests are again proxies. They may have utility if they correlate with certain desired but difficult to measure qualities, just as the 0/1 copying test did. However, IQ tests have been rightly the subject of much criticism, both for what they are testing and for whether their use is fair when individuals have different cultural or educational backgrounds.

IQ tests appeared in the early 1900s. Alfred Binet developed the first at the request of the French government, following the introduction of laws that mandated universal education from age six. The initial purpose of the test was to identify students who would not do well in a standard classroom. Binet chose to test for a variety of skills, such as the memorization of a random sequence of digits. His purpose was to predict which child would succeed in a conventional classroom. Binet himself did not believe that intelligence was a unitary phenomenon that had a quantitative measure. That interpretation arose from tests later derived from his test in the United States where, in time, IQ tests became widely used for educational placement. Some came to support the idea that IQ tests did measure something real, a so-called general intelligence. Others questioned that.

In 1994, in response to then current controversies on the nature of intelligence and the meaning of intelligence tests, the Board of Scientific Affairs of the American Psychological Association commissioned a report to make an authoritative statement on these issues as understood at that time. The report begins:

> Individuals differ from one another in their ability to understand complex ideas, to adapt effectively to the environment, to learn from experience, to engage in various forms of reasoning, to overcome obstacles by taking thought. Although these individual differences can be substantial, they are never entirely consistent: a given person's intellectual performance will vary on different occasions, in different domains, as judged by different criteria. Concepts of "intelligence" are attempts to clarify and organize this complex set of phenomena. Although considerable clarity has been achieved in some areas, no such conceptualization has yet answered all the important questions, and none commands universal assent. Indeed, when two dozen prominent theorists were recently asked to define intelligence, they gave two dozen, somewhat different, definitions.[4]

No one in the years since this paragraph was written has found a satisfactory definition of intelligence either. It is not clear even that the word has meaning as a "you-know-it-when-you-see-it" phenomenon. I have never personally found it useful for describing individuals, and I am sometimes taken aback when I hear others use it.

In the academic literature on IQ testing, the lack of any *explicit* definition of intelligence has always been evident. In 1904 the psychologist Charles Spearman published a paper entitled "General Intelligence Objectively Determined and Measured."[5] He collected data about children's school performance. He was intrigued by the high correlations of an individual's performance in different subjects, such as English, French, mathematics, and classics. From his statistical analysis of the data, he conjectured that "all branches of intellectual activity have in common one fundamental function (or group of functions), whereas the remaining or specific elements of the activity seem in every case to be wholly different from that in all the others." His "fundamental function" became the "general intelligence" or "g factor" that came to be associated with what later IQ tests were purporting to measure.

Note that Spearman's definition of this "g factor" is not explicit as far as spelling out the behavior expected of an intelligent person. It comes indirectly out of statistical analysis, namely, *factor analysis*, the technique he invented for this purpose. His definition is an implicit statistical one and not an explicit specification of behavior. The g factor he was detecting could have corresponded to parental income. In sharp contrast, the concept of educability I am advocating is defined explicitly in terms of the behavior that the capability exhibits.

The notion of intelligence has not served us well. Can we find something better?

CHAPTER 4

Some Foundation Stones

This chapter introduces the educability model in broad strokes and describes seven notions that I will use later for defining the model. The seven are *learning from examples, generalization, large memory, chaining, a Mind's Eye, symbolic names,* and *teaching.* Each is of substantial independent interest. None is unique to humans, but a particular construction based on them, which we finally arrive at in chapter 8, will be the claimed human capability for civilization.

Acquiring and Applying Belief Systems

Educability in this volume is defined as a combination of *(a) learning from experience, (b) being teachable by instruction, and (c) combining and applying theories obtained in both modes.* I shall call these the three *pillars* of educability. The aim of this volume is to give more precise meaning to these terms so as to give a detailed understanding of what the capability being defined is and what it is not. It will take several chapters, up to chapter 8, to provide this more precise meaning and explain its implications.

As already noted, humans are formally educated for extraordinary lengths of time, often more than ten years and sometimes much more. Education is concerned with handing on belief systems, a belief system being a theory often consisting of many parts created by many individuals. Those educated the longest become experts in specialized areas. A law student will spend several years learning about the laws enacted over many decades, sometimes centuries. A science student will learn about theories created and tested by many, again sometimes over centuries.

Educability enables an individual to acquire and apply a belief system that is much too broad for any one person to have had any chance of creating from personal experience. We can celebrate the abilities of the exceptional individuals who have contributed decisively to the creation and development of the most widely held belief systems. Influential scientists and writers would be examples. My focus here, however, is not on their talents. Nor is it on singular events of scientific discovery. Instead, it is on the universal human facility for absorbing belief systems created by others. The universality of this ability among humans should make it amenable to scientific understanding.

The mark of humanity is that a single individual can acquire the knowledge created by so many other individuals. It is this ability to *absorb theories at scale*, rather than the ability to contribute to their creation, that I identify as humanity's most characteristic trait. This ability would be of little use were it not for the content that the most creative individuals provide. The creative individuals themselves, however, need this more basic and universal ability to absorb previously known knowledge, on top of which they create their novel additions. Further, their novel additions would be of little use if others could not acquire them with much less effort than was needed to create them.

It would not be helpful to acquire a system of beliefs if we did not have an ability to apply those beliefs to individual situations.

In fact, it is not clear what the value of acquiring knowledge is if it does not enable us to do or understand something that we could not do or understand before. Hence *acquire and apply* is the critical notion needed with respect to belief systems.

In my view, a critical feature of education is that it can *impart knowledge that will be useful later in ways not foreseen at the time of the imparting.* In this sense, education is different from training, which imparts the skill to perform a task that *is* foreseen at the time of training. Indeed, training sessions often consist of instances of actions that exemplify the task. Education will have elements of training. The question is: What else is there in education that goes beyond training?

The educability notion offers some clarification on this. It emphasizes that a *system* of beliefs is being acquired and not just a single belief. If a learner has acquired such a system of beliefs, whether about mathematics or agriculture, then, when faced with a uniquely new situation, one not foreseen by the teacher, the learner will be able to combine several of these beliefs in a new combination in order to arrive at a novel response. The response is novel in the sense that, before that situation arose, no one, including the educator, had reason to think through what an appropriate response would be for that situation, or what the result of applying those beliefs in that situation would be.

Being educable means that one can combine pieces of knowledge gained years apart, decades later. This offers one interpretation of the words: "study as if you were to live forever."[1] A special power of being educable is that it enables one to acquire theories of considerable complexity that have many and disparate parts and may need a long time to absorb. It enables humans to master complex systems of thought, whether in science, the arts, social science, or religion.

The various parts of a belief system will typically refer to distinct aspects of the world. Acquiring these various parts so that the learner can apply them in combination is what I shall call *integrative learning*. Let's look at an example. Say you are at a spot in a strange city and want to get back to where you are staying. You need to make a plan that you have never made before. You have to use pieces of knowledge that you had acquired earlier at widely separated times, perhaps about methods of transportation, relative locations, or about other particulars of that city. More often than not, the resulting plan works, and you do not get totally lost.

Educability is more than a logical reasoning system applied to a given set of rules that one has acquired through instruction. As already noted, it also includes *a capability to learn from experience*. We are as able to learn from our own experience as we are to absorb principles others tell us. Educability, further, implies the ability to apply rules, whether learned from experience or acquired through instruction, to new situations. This ability to apply provides a potential to check out the plausibility of principles taught to us by instruction.

In each of the next seven sections, I shall focus on a separate capability that will be used as a foundation stone for the model. These capabilities are learning from examples, generalization, large memory, symbolic names, teaching, chaining, and a Mind's Eye. Each one is computational, in the sense described in chapter 1—information is being processed to meet a specification, and this specification can be realized with feasible resources. The physical substrate on which the processing takes place, whether biological or silicon, is immaterial. It is the specification of what these processes achieve that is of primary interest here and not the physical substrate. Identifying the *what* has precedence over identifying the *how*. Nevertheless, the *how* cannot be ignored. The specifications must be such that they do not require unattainable amounts of computational resources in neurons or in silicon. The number of steps and the amount of hardware in the physical substrate needed must be realistic. One cannot ascribe to human biology a computational capability that is physically unrealizable in this universe.

In the final section of this chapter, I will conclude by noting that there is no need here to model all facets of human cognition. There are many facets, such as the details of how we hear and see, that we share broadly with other animals. These may interact with educability, but they are not needed for defining it.

Learning from Examples

Hermissenda crassicornis is a brightly colored sea snail that lives in the northeastern Pacific Ocean and grows to about two inches in length. While it has a relatively simple nervous system, it can learn from experience in that training can change its behavior. For example, these snails have been trained to change their behavior in the following way when exposed to light.[2] The snail normally moves toward light to get food but will move more slowly if it expects turbulence in the water. Training it to expect turbulence after exposure to light will make the snail move more slowly toward light than it would otherwise.

The training regimen consists of episodes in which exposure to light is followed by an elevated level of turbulence in the water. The result of this training is that when food is presented, the snail will approach more slowly in the presence of light than it would otherwise. This training regimen, in which water turbulence follows light, will change the snail's neural circuits so that in the future it will respond to light differently from before.

Any kind of learning in the real world implies some level of generalization. The learned response of the sea snail occurs not only under the exactly same lighting conditions that it experienced during training, but also under other conditions that are in some way similar. For example, the light does not have to have exactly the same brightness during testing as during training. If it did, then the training would not be useful to the snail. Generalization is an essential component of any learning process. I believe that in evolutionary terms *learning from examples* and doing so with *generalization* are the most ancient components of the educability capability. More will be said about generalization in the section to follow.

In the sea snail experiment, the inputs and outputs are realized by hardware that the snails are born with. The inputs are the photoreceptive cells that detect light and the hairs that detect the motion of the water. The outputs are motor neurons that activate the snail's muscles and make it move. The input hardware I will call *sensors* and the output hardware *actuators*. The change in behavior of the snail is realized by changes in the neural circuit that connects the input hardware to the output hardware. The mechanism that makes the changes in the neural circuit I call the *learning algorithm*.

This learning algorithm is an information processing mechanism that can change the behavioral relationship between the inputs and outputs. Its physical realization is immaterial here. In the terminology of machine learning, the inputs, which here would be the photoreceptors and motion sensors, would be the *features*, and the output, the motion producing muscles, would be the *target*. In both the sea snail and a computer, there is a computational mechanism that computes the target value from the feature values. The mechanism would be a network of neurons in the snail and a computer program in the computer. The learning algorithm would change the network of neurons or computer program, so that in the future the same input values of the features would cause an action at the target different from before. In particular, if moving toward the light is painful for the snail because of turbulence, then the result of learning will be to make it more reluctant to move toward light.

The abilities of animals to learn are much more spectacular than this example for the snail. There is a rich literature on animal training experiments where what the animals are learning to detect is more abstract than the presence or absence of light. Pigeons have been trained to distinguish pictures of higher-level concepts, such as trees, water, or the presence of humans.[3] When trained to recognize trees, for example, pigeons are rewarded with food if they peck on pictures containing a tree. They are not rewarded when the picture does not contain a tree, in which case they are penalized for pecking by having to wait longer for the next opportunity for reward. After such training, pigeons do generalize well: when shown new pictures, they generally peck on those that show trees and not on pictures that do not. They demonstrate the intended behavior not only on pictures that are exactly those used during training, but also on those that are quite different. It is sufficient that the pictures of trees are recognizable by humans as representing trees of any kind, from any distance and from any viewpoint.

Similar learning capabilities have been demonstrated for artificial concepts, such as characters. Pigeons were trained to distinguish the letter "A" from the number "2," each presented in a variety of typefaces. The pigeons learned to generalize the distinction between the letter "A" and the number "2" given in new typefaces they had not previously seen.[4]

To understand the phenomenon of learning with generalization, we consider the simplest model, where the training examples are correct instances of the sought behavior. This is *supervised learning* and is illustrated in figure 1. This terminology derives from the prototypical scenario in which the learner receives a series of examples and there is a "supervisor" who provides the target value or *label* for each example. For instance, the inputs could be patterns on a screen. The label would be "A" or "2" according to which of the two symbols is depicted. The experiment in which pigeons are trained to recognize these characters is an instance of this, where the food reward during training takes the place of the supervisor-provided label.

In supervised learning, the task is to generate a *classifier*, a program that can categorize new examples. The examples are *input* to the learning algorithm and to the classifier in some format. This format is viewed as a set of *features*, which, in the case that the example is an image, would be pixels on a screen or in a camera, or the receptor cells in the retina of an eye. The *output* of the classifier is the *target*, the concept being classified, such as being a "2" or being an "A." Each example that is used for training comes with a label that identifies whether that example is a positive or negative example of the target concept. From a set of such examples and labels, the learning algorithm derives the classifier, which is a rule or an executable program. The purpose of the classifier is to predict the correct labels for new examples specified by values of the features. In this way a learning algorithm realizes the process of learning. The central mystery that needs to be resolved is how such a process can predict well for examples it has never seen before.

Before I get to that, I need to mention that the supervised learning formulation is less restrictive than the terminology might suggest. On the surface, it appears to need the presence of a "supervisor" to provide labels. In many situations, however,

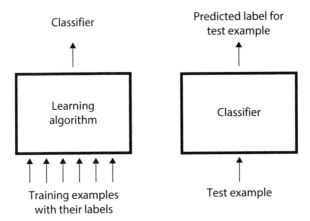

FIGURE 1. Schematic diagram of supervised learning. During training (*left*), each training example is presented with its correct label. A learning algorithm will process all these labeled examples to produce a classifier. During testing (*right*), a new example is presented, now without its label, to the classifier. The classifier will process this example and output its prediction for the label.

the world provides equivalent information without any such agent being present. Consider the sea snail example: the internal pain sensor that informs the snail of the water motion is what provides the label. The snail does not require any external supervising agent to tell it when it is in pain. This is an example of so-called *self-supervised* learning because the learner is already capable of providing the label, in the snail's case the feeling of pain, from the examples. Now the snail also wants to be able to predict the label *before* feeling the pain.

For humans, knowledge acquired earlier often provides the needed information for such self-supervision. In natural situations, there is frequently enough information available to the learner to deduce the needed label by use of prior knowledge. For example, a child at a zoo can see various interacting animals.

From previous knowledge, the child will be able to name many of the animals, foods, and behaviors. Even when no supervisor is present, the child can learn more about what each species likes to do or eat. The child's prior knowledge provides the category for each animal, food, or activity. This allows the child to learn in a self-supervised way. Seeing a chimpanzee with a banana, orange, and ball in its enclosure, a child can learn that chimpanzees eat bananas and oranges and play with balls. Knowing that oranges and bananas are fruits, the child may deduce that chimpanzees eat fruit and may conclude that they will probably also eat grapes. In a day at the zoo, the child can learn much about the nature of animals. Explicit instruction by an adult may enhance the learning, but much learning can occur without any such explicit supervision.

When an instructor *is* present to supply categories, then the choice of categories used may greatly influence what is learned. For example, through psychological experiments, Susan Gelman has found that the choice of generic noun phrases used by adults can greatly influence what children learn about the world.[5]

Generalization

The power of learning from examples is that it provides an ability to apply what has been learned to new cases. Otherwise, the process would amount to no more than memorization. What exactly are the requirements for this process of learning to be effective for generalization? What is the most that can be expected of this phenomenon, and how much of that can be achieved?

This question is the practical version of the old philosophical problem of *induction*, already discussed by Aristotle. How can we generalize beyond the specifics of the experiences we have had? How are we able to recognize a tree that we have never seen before as a tree? How do we recognize a letter "A" presented in a new font? How do pigeons recognize either?

Most likely, neither humans nor pigeons employ consciously articulated definitions for the concept of an "A" or a "tree." In the eighteenth century the philosopher David Hume famously talked about induction as based on the detection of "regularities" in our experience. We use examples that we have seen to identify these regularities. We then make inferences about new examples by determining whether they contain these identified regularities. But what exactly is a regularity? Which kinds of regularities do humans and pigeons use to learn concepts like a tree or a chair, or children use to learn the concept of fruit?

We can ask these questions in scientific terms if generalization is formulated as a computational task. Central to the notion of educability will be answers to this question of generalization provided by the *Probably Approximately Correct* or PAC framework.[6] The PAC model specifies the desired behavior and outcome of a learning process if one is to consider the process both feasible to perform and effective in achieving generalization. Probably Approximately Correct learning serves both to understand human and other biological learning as well as to characterize the technology of machine learning. Chapter 5 gives more details of this framework. It will suffice here to note that the very name already suggests aspects of its nature.

The "approximately" acknowledges that any framework for generalization needs to be tolerant of errors. In the most widely used applications of machine learning, such as speech recognition, image labeling, and language translation, as offered by internet or phone companies, it is desirable to predict accurately as often as possible, but mistakes do not generally kill anybody. Safety-critical applications are much more challenging, and it is no coincidence that the most common applications of machine learning do not currently include those. In applications like driverless cars and medical diagnosis, it may be possible to reduce error rates to better than human performance. It is not possible, however, to eliminate mistakes altogether. The "approximately" is an essential part of the generalization phenomenon, whether exhibited by humans or by machines.

"Approximately" has a compensating positive implication also, in that while accepting error is necessary, some control of the error rate will need to be achieved for generalization to be useful. Accepting unlimited error rates would make learning useless. The PAC model insists that any error rate, however small, is attainable (at least with high probability) given *enough* effort in data gathering and computation. Further, it will require that the goal of achieving lower and lower error rates can be reached with only moderately increasing, and therefore affordable, effort. At present, it is commonplace to spend enormous amounts of computer resources to learn from large datasets. The PAC framework characterizes the phenomenon that these activities exploit, namely, that the rewards of these enormous efforts are sufficient in terms of the accuracies obtained to make the efforts worth making.

This explains the meaning of "approximately" as far as the nature of the guarantees that PAC learning provides. How about the "probably"? "Probably" is a qualification of the guarantees of efficient error control just described. These guarantees are delivered only with high probability and not with certainty. We can be unlucky and have unrepresentative training examples that are not characteristic of what is to be learned. Although the probability of getting such a spurious training set may be small, if we are unfortunate enough to have such a bad training set, then low predictive accuracy may follow.

Large Memory

A deceptively simple question is: What is the memory capacity of the human brain? The *Oxford English Dictionary* has about six hundred thousand words, but Shakespeare used only about twenty-five thousand distinct words in all his plays. What is the largest number of words that any one human can use or can recognize? We can ask analogous questions for every area of human expertise. How many diseases does a primary care doctor need to be able to diagnose? How many chess opening lines does a grandmaster need to know?

Some have made experimental estimates of human memory capacity. In 1973 Lionel Standing showed subjects up to ten thousand pictures of natural scenes.[7] Some days later, he showed the same subjects pairs of pictures, one previously seen and one not. They were asked to identify which one they had seen before. The error rate generally went up with the number of pictures shown, reaching 27 percent error at ten thousand.

In recent years memorization has become a sport with standardized events, competitions, and world rankings. At the time of writing, the world record score for memorizing sequences of decimal digits was 630 when given five minutes for the memorization and 4,620 when given an hour.[8] Here the scores correspond to the number of digits recalled, with certain penalties for mistakes.

Researchers in animal behavior have also investigated memory capacity in other species. Squirrels cache many nuts each year. Birds cache even larger numbers of seeds. It is believed that both species can recall a significant fraction of the thousands of specific hiding locations they use each year.[9]

Scene memorization by animals, in the sense of Standing's experiment, has also been explored. In one experiment, baboons could process over 5,000 pictures and still recall a significant fraction, while pigeons could do that for up to 1,200 images.[10]

Do large brains have any useful purpose beyond the memorization of large numbers of individual memories, such as the hiding places of nuts, landmarks for navigation, or where you left your keys? The answer must be yes. We use our memory also to store general learned information. This information may be both what we have learned from personal experience as well as what we have been taught by instruction. There is an analogy here with general-purpose computers. Computers, including the ones we have in our cell phones, now have very large memories. Computers use memory for storing both programs and data. In human terms, these correspond, respectively, to the general rules we know and memories of individual instances. For example, a general rule is that to find my keys, I search in my pockets. An individual instance would be the memory of having put them in a specific unusual place, say, on a chair.

As already mentioned, a most spectacular development in human evolution has been the tripling in brain size since our ancestors and those of the other living apes parted company. Educability offers a suggestion for why this might be useful. We do not need it for memorizing more hiding places for nuts. We use it for storing information to help decide what to do or think next in our lives.

Symbolic Names

Some neurons in biological systems are sensors that respond directly to an external physical stimulus, such as movement or light. Other neurons are actuators, producing a physical result such as the motion of a limb. As we saw earlier, even simple creatures like the sea snail need and have both. For both sensors and actuators, we can think of the neurons that correspond to them as representing some physical reality, such as light, outside of the neural system. These are physical representations because the activity of the neurons corresponds directly to some physical reality outside of the neurons. The physicality of the representation derives from the correspondence between the internal activity of the neurons and the outside reality, for example, between a photosensitive cell being stimulated and the light that stimulates it. When one is learning a classifier, the physical sensors can serve as features or inputs and the actuators as targets or outputs.

In the life of the mind, we humans use representations that are far removed from such physical sensors and actuators. We can represent whatever concepts life calls on us to represent. It is widely recognized that the use of symbols is an important characteristic of human cognition. The problem is that it is not so simple to characterize what this notion means or how our brains process symbols.

Does a dog that answers to the name "Socrates" have a symbolic name for itself? As a working criterion, an entity will be said here to use A as a *symbolic name* for B if (i) there is arbitrariness in the choice of A in the sense that many alternatives would have worked also; (ii) the entity will be reminded of A when given B; and (iii) the entity will be reminded of B when given A.

This definition misses out some secondary characteristics, most notably that for symbolic naming to be useful, the A needs to be simpler than B. For example, a short word is a useful name for a complicated concept, but not vice versa.

How about the dog that answers to the name Socrates? From the owner's perspective, this is symbolic naming since (i) the owner had chosen the name arbitrarily; (ii) when thinking of the dog the owner will be reminded of the dog's name, Socrates;

and (iii) the owner is reminded of the dog when given the word Socrates.

From the viewpoint of the dog Socrates, things are not so clear. The name is still arbitrarily chosen. But on hearing its name, does it think of itself, rather than just react in some trained fashion? When thinking of itself, does it think of the word "Socrates"?

Symbolic names in the sense just defined are essential in human mental life. If we talk to someone about a third person who is not present, it is certainly more convenient to refer to their name than to give a convoluted description of who they are in other terms. Scientists often invent new words to describe complicated new concepts. "Transistor" and "laser" are examples. Such names may be acronyms of other words or have other indirect connections with the meaning, but many other names would have worked equally well to refer to these concepts.

A generic capability for providing a symbolic name to an arbitrary concept conveys enormous freedom not available to organisms whose learned classifiers can have only physical sensors as inputs and physical actuators as outputs. You can make up any sequence of sounds that you can utter, call your dog that often enough, and it will work as a symbolic name, at least for you. You can do the same with the names of new concepts in mathematics or physics, or in naming new species of birds. You *can* organize an expanding field of knowledge using an expanding set of symbolic names and *cannot* without them.

Chapter 8 will discuss how the realization of symbolic names, in neurons or in silicon, raises a host of problems. The primary one is determining where and how the information that the name refers to is to be stored. In a physical printed dictionary, you store names in their alphabetical order: Socrates's information would be under "S." In a digital computer, there are absolute addresses, which are numbers, and you would need to associate the sequence of letters in "Socrates" with one of them. In the mammalian brain, as far as we know, there is no numerical ordering or addressing of the neurons, and the problem is more of a challenge. It is perhaps not surprising that it took a long while for evolution to solve this problem of symbolic naming.

I regard the assignment of an arbitrary symbolic name to a concept as fundamental to human cognition and to educability. Many discussions of symbols jump at once to the more arcane issues that arise in human language: Where do the complex recursive grammatical structures that occur in natural languages like English come from? What is the order among the subject, verb, and object in a sentence in various languages? Each natural language among the six thousand or so currently spoken incorporates vast richness and complexity. Fortunately, all this complexity is peripheral to the discussion needed here. For educability, the most critical contribution of language is that of supplying words, which are the means of assigning symbolic names to concepts.

Language does have a second key role, that of providing a way of describing theories so that they can be conveyed explicitly from teacher to student. The evolution of language may have been a prerequisite for the emergence of civilization by virtue of these two roles; however, I do not believe that language by itself explains, or that its detailed structure is central to, the creation of civilization. Language as such is not the sought-after Civilization Enabler.

As mentioned earlier, symbolic naming is not so trivial to implement in brains. Do nonhuman animals respond to arbitrary names? You can call your dog "Socrates" and it will give the appearance that it understands that the name refers to itself, but this does not imply that the dog has a general symbolic naming capability. Even the limited ability that dogs have in this regard is apparently rare. Which animals besides humans and dogs respond in this way to an arbitrary name they are given? A recent study suggests that cats have some ability to respond to their own names over other similar-sounding words.[11] Bottlenose dolphins have vocal identity signals that they make, which differentiate among them. These dolphins respond to hearing their own identity signal. They occasionally make the identity signal of another dolphin as if to call them.[12] Some bird species, such as parrots, also respond to their name, but this behavior does not appear to be widespread in the animal kingdom beyond these groups. This limitation may be due to the difficulty of assigning arbitrary names in the brain architectures that have evolved on Earth.

In this work, I use "naming" for a general capability of naming arbitrary concepts. Recognizing one's own name, as discussed in the previous paragraph, is just a special case. There is evidence that some primates and birds *can* be taught a substantial vocabulary of words and their real-world meanings. This requires more than merely responding to one's individual name. It requires that for each word the animal behaves appropriately to its meaning. A chimpanzee named Nim Chimpsky reportedly learned 125 signs of American Sign Language in forty-four months.[13] A gorilla named Koko acquired an active vocabulary of a thousand signs from a modified version of American Sign Language and could recognize two thousand words of spoken English.[14] A border collie called Chaser was reported to be able to distinguish over a thousand dolls by their name.[15] While these reports have attracted differing interpretations, taken at face value they suggest that large vocabularies, with at least one-directional associations between words and their meanings, are not unique to humans.

To summarize, symbolic naming as defined here is the ability to learn to associate in both directions an arbitrary name—which may be a snippet of sound or other information—with something else. Educability will need full use of symbolic naming, integrated with the other specified capabilities. Let us turn to teaching next.

Teaching

In supervised learning, as defined earlier, the learner's environment is truthful—it provides true examples and true labels. Beyond that, it seeks neither to help nor to hinder the learner. In that sense, it is *learning neutral*. Teaching is different. Now, an external agent seeks to help the learner learn more effectively than they could if left to their own devices. Teaching is an ever-present feature of human learning. A teacher is usually present in formal settings of education. Even when one is learning by oneself, the material used has usually been prepared for teaching. Encyclopedias contain carefully curated material presented to facilitate self-education. Those who contribute to encyclopedias are teaching. Textbooks and online courses certainly teach. Teaching apparently occurs in every human culture.

Is there any evidence of teaching in the nonhuman world? Teaching had long evaded observation in other species, and many believed that it did not exist. In recent decades, the balance of opinion has swung toward accepting that teaching occurs in many species.

Researchers in animal behavior have sought to define what constitutes teaching. A young animal may be learning in the presence of an older one, but this could be by imitation without any intentional teaching. On the other hand, we cannot define teaching in terms of an intention to teach, since we are in no position to assign intentionality to animals. Is there a useful criterion by which to recognize teaching? One suggestion is that three requirements need to be satisfied: (i) a changed behavior of the purported teacher when in the presence of the purported learner; (ii) a cost to the teacher for this behavior; and (iii) evidence that the learner has gained more knowledge or skills after experiencing the teacher's changed behavior than it would have otherwise.[16]

In the past twenty years, several species have been observed to exhibit behaviors that satisfy this definition of teaching. Chimpanzees hand sticks to their young for poking into termite nests to help them retrieve termites. This satisfies the three requirements as this behavior is seen only between mothers and their children, it costs the mother not getting the termites for herself, and it improves the termite gathering of the offspring.[17] A certain kind of ant, the *temnothorax albipennis*, shows its companion ants the way to a new food source, slowing down as necessary to allow the learners to keep up.[18] Meerkats, members of the mongoose family, feed on live scorpions. Adults start feeding their young dead scorpions. As their young get older, they are fed with live scorpions that have been rendered unable to sting. As time goes on, the young are fed scorpions that are more and more capable of stinging, since it is live scorpions that the meerkats need to learn to catch.[19] Golden lion tamarins, a kind of monkey, find prey in tree hollows. Sometimes the tamarins do not reach into the hollows themselves. Rather, they call their young over and allow them to reach in and get the food for themselves.[20]

These examples suggest that teaching goes back a long way in evolution and appears in many branches of the animal kingdom. Since teaching is a powerful method for improving skills, it would be surprising if evolution had failed to discover it early.

Among humans, teaching styles vary from culture to culture. In Zen Buddhism a *koan* is some prose, often paradoxical or ambiguous, given to students to test their progress. The following is an example:

> Zen teachers train their young pupils to express themselves. Two Zen temples each had a child protégé. One child, going to obtain vegetables each morning, would meet the other on the way. "Where are you going?" asked the one. "I am going wherever my feet go," the other responded. This reply puzzled the first child who went to his teacher for help. "Tomorrow morning," the teacher told him, "when you meet that little fellow, ask him the same question. He will give you the same answer, and then you ask him: 'Suppose you have no feet, then where are you going?' That will fix him."
>
> The children met again the following morning. "Where are you going?" asked the first child. "I am going wherever the wind blows," answered the other. This again nonplussed the youngster, who took his defeat to the teacher. "Ask him where he is going if there is no wind," suggested the teacher.
>
> The next day the children met a third time. "Where are you going?" asked the first child. "I am going to the market to buy vegetables," the other replied.[21]

In the framework of educability, this piece of prose would correspond to an example given to a student by a teacher. On the basis of previous knowledge, the student would draw conclusions about the circumstances described that are not explicitly stated. Hence a teacher can be helpful to a student who is learning by example. In using the koan, the teacher is helping via the simple act of pointing to an instructive example.

As will be detailed later, educability also encompasses, and crucially so, the more powerful method of interaction where the teacher explicitly describes the rule and ultimately a whole belief system that the learner is expected to absorb and apply.

There are intermediate ways also in which a teacher can help. A teacher can just name a concept new to the learner and then provide examples, so that the student can learn that concept from examples. Such new concepts may later become features for learning further concepts. At some point we first heard about Covid-19. Now we can learn other concepts in terms of it, such as Covid-19 vaccine.

As we have seen, teaching is not unique to humans. In educability we are particularly interested in the specific form of teaching in which the teacher communicates rules, theories, or instructions explicitly. We would expect that language is important for this, but it is not entirely essential unless the rules involve symbolic names. Certain animal behavior experiments described below shed light on the question of teaching without the use of language.

A much-studied area of animal learning is *social learning*, which means learning from others (who may or may not have the intention to teach). In one such experiment, the experimenter demonstrates a solution to a physical problem posed by a gadget and expects the subject to be able to solve the same problem subsequently. One instance of such a gadget is a transparent tube with stoppers at both ends and a food reward inside. The stopper at one end is fixed but has a hole in it. The stopper at the other end is removable if pushed out from inside. The experimenter demonstrates to the subject that they can retrieve the food reward by pushing a rod through the hole in the fixed stopper, thereby pushing out the stopper at the other end. In that experiment, it was reported that two-and-a-half-year-old humans successfully solved this problem, but chimpanzees and orangutans of any age had difficulty.[22]

Solving such a task is like acquiring a theory by being taught through instruction. While we normally associate theory teaching among humans as involving language, a physical demonstration as in the experiment just discussed is largely equivalent. This suggests that the capability of being taught through instruction may be tested independently of language.

Note that the ability of animals to solve problems of this nature depends subtly on how the problem is presented. It is possible that there is some criterion of task difficulty at which human and chimpanzee abilities diverge, but such a criterion remains to be identified.

A distinction made by some researchers in animal behavior is between *imitation* and *emulation*. Imitation here means repeating step-by-step what the teacher has demonstrated. Emulation means solving the problem to which the teacher has shown a solution, possibly by a different, perhaps simpler, method. The distinction between the two can be highlighted by having the demonstrated solution include some obviously redundant steps—maybe a redundant move is made in the transparent tube example that is immediately reversed. The imitator will repeat the obviously redundant step while the emulator will omit it. If the essence of being human were superior problem-solving ability, then we humans would be the emulators and the chimpanzees the imitators. Perhaps unexpectedly, there are experiments reported that show a converse tendency—it is humans who repeat obviously redundant steps and not chimpanzees.[23]

These experiments also showed that chimpanzees can learn to imitate like humans if brought up in a human-like environment. This is consistent with the idea that humans acquire theories from a teacher, as if by rote learning, recognizing that a teacher is someone special to be imitated. Chimpanzees not brought up in a human-like environment may be behaving more as if the teacher was just another phenomenon in the world, and not one created to help them learn.

Interestingly, some tension has been demonstrated in humans between accepting what a teacher says and learning for oneself. In experiments on preschoolers, the presence of a teacher explaining the purpose of a toy has been shown to have the effect of reducing the amount the children subsequently discover about the toy through their own experimentation.[24]

Chaining

The cleverness of the Belfast chimpanzees in finding a method of escape, described in chapter 1, is evidence of their abilities to plan and reason. While each component of the escape may be commonplace as a separate behavior, such as climbing along a branch and jumping to grab something, there needs to be a place where two constituents that in combination make for a novel action are brought together for the first time. This bringing together of disparate pieces of knowledge that fit together is *chaining*. It is a form of reasoning.

Such chaining of separately learned behaviors has been long studied in animals. One context is that of *Classical* or *Pavlovian conditioning*. Here one trains an animal to associate an involuntary response (one not under conscious control) with a stimulus arbitrarily selected by the experimenter. This now celebrated phenomenon is named after Ivan Pavlov, a physiologist who was awarded a Nobel Prize for his earlier work on digestion. Normally a dog will salivate when receiving food (an involuntary response). Pavlov trained a dog to salivate on hearing a bell (the arbitrary stimulus). He did this by a regimen of first ringing a bell and a little later giving the dog some food. After this regimen had been repeated long enough, on hearing the bell the dog would salivate immediately, *before* obtaining any food.

This phenomenon of an animal acquiring such a chosen association has been widely demonstrated. The sea snail experiment described earlier is another example. The phenomenon can be found in many species and across many choices of stimuli (e.g., bell, light) and responses (salivation, slowing down). As further examples, almost any stimulus paired with an air-puff to the eye or a mild electrical shock can elicit the responses of blinking or twitching.

Pavlov also showed that associations can be chained. In that experiment, he first trained the dog as before to salivate on hearing a sound. Now as a separate step, a black square was presented to the dog at the same time as the sound, and without the food. The result was that after some training, the black square becomes sufficient to elicit salivation, even without the sound and without the food. This is called *second-order conditioning*. It is a basic form of chaining in which associations learned at distinct times are combined.[25]

Chaining is therefore yet another capability that goes far back in evolution.[26] For educability, we shall need a particularly flexible version of it. The Mind's Eye will be a device where rules learned at separate times can be brought together and chained so as to produce a compound response to a complex, novel situation. We turn to this next.

A Mind's Eye

The Mind's Eye performs another vary basic function, that of scene analysis. This enables distinct parts of a scene to be distinguished as separate objects. For instance, the Zen story told earlier is quite complex, describing as it does three protagonists in different and changing states of mind. In processing the story, one needs to consider the changing viewpoints of each of the participants in turn. Such an ability to change focus between different parts of a scene is sometimes called *attention*.

The Mind's Eye encapsulates mechanisms for such scene analysis, detailing the individuals in a scene and the relationships among them. In a scene containing the four qualities of *big, toy, tiger,* and *duck,* detecting the presence of these attributes may not be sufficient. One may want to distinguish between being confronted by a big tiger and toy duck, as opposed to a toy tiger and big duck.

Being able to recognize individual elements in a scene is useful, but recognizing multiple objects and the relationships among them is even more powerful. Recognizing the presence of an elephant in a field or the attributes of individual elephants is useful. Recognizing that there is a herd of elephants and that the adults are protective of their young provides information beyond what can be expressed if only a single object can be considered at a time.

Cognitive scientists call *working memory* the mechanisms of the brain that relate most directly to what one is paying attention to or is aware of at a given time. The Mind's Eye, as we use the phrase here, is closely related to this notion. It models the computational task of representing a situation and bringing together the information that is relevant to it from various parts of long-term memory. A novel situation may prompt a Mind's Eye to bring together a combination of rules that has never been brought together before.

The psychologist George Miller wrote that humans are able to hold at most *seven plus or minus two* concepts in their working memory.[27] I believe that inherent computational limitations impose the cognitive limitation that Miller noted.[28] The formalization of the Mind's Eye that will come later will be correspondingly economical in its use of short-term memory.

Reasoning has taken pride of place in human self-regard. There have been extensive investigations on the question whether animals can reason also. In one famous experiment, chimpanzees were able to solve the problem of getting to bananas suspended from a high ceiling in a room where several boxes were present. The chimpanzees figured out that they could stack the boxes and then climb on top of them to reach the bananas.[29]

Assuming that neither evolution nor previous learning experiences had primed the chimpanzees to solve this problem, one has to conclude that the animals composed in some part of their brain the sequence of actions they later executed. The Mind's Eye is our formulation of what that brain part does. From these and other animal experiments, we see that chaining for solving planning and reasoning tasks was present long before the arrival of humans. Educability will need to use these elements in a flexible way. The roles of the Mind's Eye are both to represent the several parts of a scene and to chain multiple pieces of knowledge on that representation.

The computational limitations alluded to earlier on the number of concepts that can be kept in mind at any one time imply that for the Mind's Eye to function, it will need to use *abstractions*. For example, the abstraction "herd" will enable it to process a scene with many elephants without needing to keep track of all the individual elephants.

The Mind's Eye is also the locus of imagination. It can represent and analyze situations constructed in the mind, not just those physically witnessed.

The Rest of Cognition

I have now introduced seven foundation stones that are important for defining educability. Human behavior and cognition have many facets beyond these. Educability is proposed as an abstraction for the most characteristic aspect of the human mind, but not for its entirety. Many of the omitted facets are relevant to educability even if they are not needed for defining it. I will therefore acknowledge here some of these other facets.

Unsupervised learning is a general term used to describe a variety of learning phenomena where no supervisor is present. Some authors classify what I called self-supervised learning earlier as unsupervised, but, as we have seen, it can be understood in terms of supervised learning. Detecting correlations, or co-occurrences of events that happen with much higher frequency than they would if they were independent, is unsupervised learning in a more fundamental sense.

Reinforcement learning refers to situations in which the learner interacts with the world and different actions may lead to different benefits. Playing games like chess exemplify such situations. Generalization is required just like in supervised learning, but the states produced by the learner's actions, such as the chess game positions, become further examples. Often the labels, such as whether a game has been won or lost, are received with much delay following the actions or moves.

We can also extend supervised learning in various other ways. One direction is to allow the learner to choose the examples and to ask a teacher what the correct label or classification is. This is *learning with membership queries* (where the word "membership" refers to the class of examples that have a common label). This contrasts with the basic model in which the world presents random examples from the natural distribution of occurrences that are found in the world and the learner cannot influence the choice of examples. By asking such a membership query, the learner can choose a question to which an answer would be particularly informative given the learner's state of knowledge. When hearing a presentation, humans often itch to ask questions. This suggests that getting answers to questions *we* have posed might make learning more efficient as compared to passively listening to someone else or observing the examples that the world has chosen to provide to us.

Humans and many other species *play* and *explore*, especially when young. When two-year-old humans are in an object-rich environment, they make extraordinary efforts to try everything out, poking, testing, and modifying the objects around them. These behaviors can be viewed as membership query learning. Instead of passively waiting for examples, the child is choosing them actively, possibly those that would yield the most useful new information. They find out whether an object of interest bends, bounces, or is soft. No supervisor is present. The world provides the answers directly and acts as a surrogate teacher.

Such query learning or experimentation is a strengthening of the basic model of supervised learning. In the opposite direction, there are ways of making the model weaker. Perhaps the simplest is the *memorization* of an *instance*. Squirrels memorizing where they hide their nuts is instance learning, where there is no need for sophisticated generalization. The same holds for memorizing a picture, a face, or a fact.

The notion of memorization makes sense only if one defines what operations on the memories will be required. If there are no operations, then there is no benefit to memorizing. Among useful operations on memorized information, one is *recognition*. This is the task where one is presented with some input and has to recall whether it has been seen before. Another operation is *pattern completion*, in which one is given partial information about something, such as an object partially hidden behind another object, and the task is to reconstruct the rest. A further one is *association*, in which certain information is to be recalled when prompted with a certain different cue.

Humans can often memorize something after seeing it just once. This is "one-shot" or "one-trial" learning. Nonhuman animals can clearly do this too. The second chimpanzee that escaped from its enclosure in Belfast Zoo in the incident described earlier presumably used one-trial learning.

Some authors find contradiction between PAC learning, which requires many examples, and one-shot learning. There is no contradiction if one recognizes them as being different phenomena. PAC learning allows finer and finer distinctions to be made between concepts by finding generalizations with smaller and smaller error at the cost of larger and larger effort. With one-shot learning, the ability to generalize correctly is more restricted and needs to be seen some other way.

My view is that at any instant an individual has some measure of similarity among objects, perhaps partly acquired through evolution and partly learned. When presented with an object and deciding whether it is the same as one seen the previous day, this *similarity metric* will tell us whether the two objects are to be interpreted as the same or not. A squirrel may be using some preconceived criterion that if two hiding places are close enough, then they are hiding the same nut.

As another example, suppose you see a person one day selling fruit at a market stand, and the following day you see a similarly dressed person in the same place. Your basic similarity metric will give you an opinion as to whether the two people are the same. If the two people happen to be identical twins, then you may make a mistake.

PAC learning is a stronger statement about making distinctions. It asserts that for two different concepts, given enough examples and enough computation, you will be able to learn to distinguish them better and better the more effort you make (so long as there is a distinguishing criterion that is learnable). In the metaphor of identical twins, if you make enough of an effort to get to know them, then eventually you *will* learn to distinguish them, even if your initial similarity metric did not. In short, PAC learning is concerned with distinguishing concepts where the effort needed may go up as the concepts get more and more difficult to distinguish. In contrast, when doing recognition after one-shot learning, you apply the similarity metric you happen to have at the time.

Any cognitive entity, whether a person or a machine, will be forced to make decisions all the time about whether a given pair of objects is the same or different. Hence any such system, good or bad, will have some similarity metric. It is in no way surprising that in the natural visual world we can do a reasonable job at recognition after one-shot learning. This is an essential task for which evolution has prepared us well. This phenomenon, however, is different from learning with generalization, or PAC learning, for which evolution has prepared us too.

Last, will there be room for novelty and imagination in our system? The answer is *yes*. The part of educability that enables the transfer of explicit rules between individuals is powerful but by itself does not produce anything new. Since novelty is central to the most valued creations of humanity, whether in science or literature, one needs some account of how it can arise. I shall return to this in chapter 6.

Before we use the foundation stones described in this chapter to build our model, we shall go on a detour to explain and justify the computational approach. Understanding the nature of human mental faculties is one of the great scientific challenges of our time. I believe that computation offers a viable path to addressing it.

CHAPTER 5

Computation for Describing Natural Phenomena

Turing Computation

Alan Turing defined our present-day notion of computation in the 1930s. He gave a mathematical definition, soon to be called a Turing Machine, intended to capture *all* ways of processing information that could be considered to mechanically follow rules.

A Turing Machine can be viewed both as a computer program and as a machine that executes the program. It provides perhaps the simplest way of describing a mechanical procedure. An instance of a Turing Machine has an infinite storage *tape*, which at any time would have a finite number of symbols written contiguously on it. It has a *head* that at any time is looking at one place on the tape. At any time, the machine is in one of a finite number of *states*. There is a finite set of *rules* that control the head. At any time one rule is executed. The choice of rule depends on the symbol then under the head and the current state of the machine. The rule can change the symbol under the head, change the state, and move the head left or right or not at all. The input to the machine is a sequence of symbols on the tape. The output is the contents of the tape if and when a special so-called *final* state is reached, from which no further rules apply, meaning that the process stops. Some computations never reach this final state and go on forever. The reader can look up a more extended discussion elsewhere.[1] The main point is that Turing's definition is simple and unambiguous.

As far as the question of whether a Turing Machine can be built in practice in this world, the answer is, obviously, yes. One can make a physical device that performs the specified steps of reading and moving along a tape. If the Turing Machine reaches its special final state after making a finite number of steps, then the process is finite. There is no reasonable way of disputing that a Turing Machine computation that stops for all inputs constitutes a finite process, since in all cases it performs a finite number of pedestrian mechanical steps.

On the other hand, but much less obviously, it has become generally accepted, for reasons I will come to later that I shall refer to as *model robustness*, that everything that can be considered to satisfy the condition of being a finite process *is* computable by some Turing Machine. Turing Machines have the same power regarding what they can compute as present-day digital computers.

Beyond defining what a process was, Turing also showed that there existed unambiguously defined questions for which, indeed, *no* process existed. Being able to define a mathematical question, such as that of distinguishing between equations that have a solution from those that do not, is therefore not equivalent to there being a process for making that distinction in all cases. In particular, for a certain class of equations there *provably* is no process that determines for every equation whether or not it has a solution.

Turing had succeeded in characterizing a very important phenomenon, which we now call *computation*, in a precise mathematical way. The remarkable thing is that while his characterization is simple—I summarized it above in a single paragraph—the phenomenon it describes is pervasive. In our present world, computation has become ubiquitous in our technology, with data centers, supercomputers and the billions of cell phones around us all incessantly computing away. The passing on of information across generations in the form of DNA can be viewed as computational also, involving as it does familiar finite operations such as gene copying, error correction, and randomization. As already discussed, the operations brains perform may also be viewed as computational.

Probably Approximately Correct Learning

In the early 1980s I became intrigued by the phenomenon of learning. Millions of children every day hear words used in various natural situations and somehow learn the intended meanings. We were all shown different instances of chairs when growing up, yet we agree to a remarkable extent on what is and what is not a chair. The surprise is the extent to which we agree about chairs, given that we grew up seeing different ones. The fact that there are strange borderline cases where reasonable people disagree is only to be expected.

Millions of children learning the meaning of words from different instances, and then largely agreeing on what they mean, is a reproducible natural phenomenon. There should be a scientific explanation for how it is possible.

I was equally impressed by the fact that this phenomenon seems to have severe limits. While chairs have some regularities or properties in common that enable us to learn the concept quite easily, there are many regularities in the world that we find more difficult to detect. That hygiene gives protection against disease is now well-accepted, but this regularity escaped general recognition for millennia.

This dichotomy between feasible-to-detect and infeasible-to-detect regularities, led me to the observation that if the mystery of human learning was to be resolved, the resolution would have to distinguish between patterns that could be learned routinely and those that were computationally too difficult to learn. The outcome of these considerations was the Probably Approximately Correct model of learning mentioned earlier. The PAC model is a characterization of what one can expect a learning algorithm to achieve in the realm of feasible-to-detect regularities.

In thinking about this problem, I became aware that to solve it one would need to solve what philosophers had called the problem of induction, to which I have also referred. Since antiquity, it has been recognized that human thinking involves more than making deductions from facts. There is the important other component of generalizing from experience. On what basis can we make predictions about situations we have not experienced from situations that we have experienced? Hume articulated this conundrum in *A Treatise of Human Nature* as "there can be no demonstrative arguments to prove, that those instances, of which we have had no experience, resemble those, of which we have had experience." He provided the clue that "probability is founded on the presumption of a resemblance betwixt those objects, of which we have had experience, and those, of which we have had none."

By the 1980s the fields of probability and statistics had indeed developed methods of inference, but none of these methods fit the bill for distinguishing between what is *computationally feasible* and what is not with regard to learning. It seemed that this problem was not understood even in principle. I believed that what was needed was a formulation that was general enough to address the philosophical problem of induction and was at the same time capable of distinguishing those kinds of generalizations that were computationally feasible from those that were not.

By then developments in computer science had highlighted that one can capture computational feasibility by the criterion of *polynomial time computation*, namely, the requirement that computation time, the number of basic steps, should grow as a fixed power, such as the square, as opposed to an exponential, of the *size* of the input. When using the polynomial time criterion there is always a numerical parameter, say n, that measures the input size. This size is sometimes chosen as the number of characters needed to describe the input, or, in learning tasks, the number of training examples.

The polynomial time criterion distinguishes between computational costs that grow as n^c rather than as c^n when n grows but c is a fixed constant number. For example, consider an 8×8 chessboard. Let $n = 64$ be the number of squares. The number of ways of placing the white and black kings on different squares is $64 \times 63 < 64^2$, since there are 64 choices for the former, and then 63 for the latter. This is *polynomial growth*, the affordable kind, less than n^c. Here $c = 2$, the number of pieces we are placing, and n the total number of squares on the chessboard.

On the other hand, suppose we take the same 8×8 chessboard and, proverbially, place 2 grains of wheat on the first square, $2^2 = 4$ grains on the second, $2^3 = 8$ on the third, etc., until 2^{64} grains are placed on the last square. Then a calculation shows that the total amount of wheat needed would amount to about a thousand times the current annual world production of wheat. This is unaffordable *exponential* growth c^n, for the same $n = 64$ and $c = 2$.

PAC learning captures the phenomenon of affordable, or *computationally feasible*, generalization in terms of this polynomial time notion. It does so in terms of supervised learning, which we have already discussed in chapter 4. Here a learner gets labeled examples of a hidden criterion and seeks to generalize from this knowledge to examples not previously seen. The learner needs to derive a *classifier* from the labeled examples. Such a classifier is a procedure for labeling new examples in terms of the hidden criterion, such as whether an image contains a chair. For an image x the *label* will have value "1" if the image represents a chair and "0" if it does not. (In general, "1" will represent positive examples and "0" negative ones.) As in figure 1 in chapter 4, the output of the classifier on a new example is the prediction of the label for that example.

The supervised learning task consists of two phases. In the first, the *training phase*, the learner is given a number of examples and also the correct label for each. In the second, the *testing phase*, the learner is given a further example, but without the label, and the learner's task is to predict the correct 0 or 1 value of the label for that new example. Does a new image depict a chair, or does it not?

The learner uses a *learning algorithm* to derive the classifier from the labeled examples. What should we expect of a learning algorithm before we declare it successful? The PAC model spells out some requirements. On the one hand, these requirements need to be onerous enough that, if achieved, the algorithm would be clearly realizing effective learning. On the other, the requirements cannot be so onerous that they are not achievable in practice.

In any rich enough setting, such as realistic photographs of furniture in actual rooms, *some* tolerance for errors is essential. This is because it is always possible that a test example is of a rare enough kind that it is unlikely that any in the training set resembles it. For that rare type, the learner would have no valid basis to make a prediction. Some errors, therefore, must be tolerated.

If, on the other hand, errors are tolerated with no limit, then the learning will be of no use. Hence for the learning process to be useful, one needs to *tolerate but control* the error rate.

For this reason, a big component of the definition of PAC learning is about *controlling* error. Informally, PAC learning stipulates that with high probability the learner will label each new example with *small enough probability of error* after investing only *practically feasible effort* in processing the examples during training and testing.

For feasibility, we insist that the computational effort, or number of steps taken, be polynomial bounded, in the sense previously discussed. To make this concrete, one needs to quantify the total effort made by the learner. The simplest option is to consider the number of computational steps made by the learner. Calling this number N, one asks how the errors decrease as the effort N increases. Note that it takes at least one step to inspect each training example; hence N will always at least equal the number of examples that the learner inspects. In most useful learning algorithms, the number of steps N will far exceed the number of examples. If we measure the input size n as the number of training examples, then for processing a dataset of $n = 1,000 = 10^3$ examples, one may take, say, $N = 1,000,000 = 10^6$ computational steps. This is the square of the number of examples and would be consistent with a polynomial time learning algorithm where the runtime is $N = n^2 = 1,000,000$ for the case

$n = 1,000$ examples. This quantity N is a basic measure of the resources used in the learning. Computation time, electric power, dollars, and any consequent carbon dioxide emissions all depend on this N.

A fixed learning algorithm applied to a fixed dataset will achieve a certain error in mislabeling new examples. Consider again the case that the effort made is $N = 10^6$ steps. Suppose that the error achieved is 0.1 percent, meaning that we expect to make a mistake in predicting whether an image contains a chair once in 1,000 trials. We should demand a meaningful reduction of the error if we put in significantly more effort, say four times more or $N = 4 \times 10^6$ steps. The definition of PAC learning requires that the error go down fast enough that the reward is worth the extra effort. In particular, it requires that the error diminish as a fixed power of $1/N$, so-called *algebraic* or *power law* decay. For example, this power could be $1/2$. In that case, the error would decrease proportionately with $1/N^{1/2}$. For the example above, if the error is this $1/N^{1/2}$, then when $N = 10^6$ the error *would* be $1/(10^6)^{1/2} = 1/(10^3)$ or 0.1 percent, as claimed. The point is that for four times more effort, namely, $N = 4 \times 10^6$, the error would go down to $1/(4 \times 10^6)^{1/2} = 1/(2 \times 10^3)$ or 0.05 percent, meaning that we could expect to make a mistake once in 2,000 trials, rather than the once in 1,000 trials previously seen. We would be halving the error by making four times the effort.

In general, PAC learning requires that whatever effort we make, we can further reduce the error by any fixed fraction by increasing the effort by *some* corresponding fixed multiple. In the example above, the fractional reduction of $\frac{1}{2}$ in the error is achieved with 4 times the effort. If for the same error reduction of $\frac{1}{2}$ an increased effort by a multiple of 10 were needed, that would still suffice for the definition of PAC learning. However,

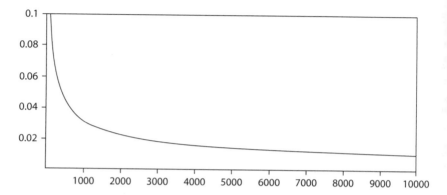

FIGURE 2. An example of algebraic decay. For effort N ranging from 100 to 10,000 along the horizontal axis, the curve shows how the error, here $1/N^{\frac{1}{2}}$, decays from 0.1 to 0.01, as shown on the vertical axis.

if the multiple grew as N grew, rather than being a fixed constant value such as 10 for all values of N, then the rewards of increasing the effort by a fixed multiple could become vanishingly small for large N.

The point is that PAC learning insists that the error control be good enough that one obtains adequate rewards for any extra effort by way of more examples and computation. The definition of PAC learning therefore captures an optimistic wish, which, if realized, would achieve effective and feasible learning with generalization. It turns out, fortunately, that this wish is not *overly* optimistic. PAC learning is achievable in some settings. For some simple learning algorithms, it can be mathematically *proved* that they achieve the efficient error control just described.[2] For some others, the error control is currently beyond mathematical analysis, but experiments show that this error control is achieved for useful data sources.[3] The data centers of large internet companies can burn up millions of kilowatt hours of electricity and millions of dollars' worth of computing time when learning from large datasets.[4] Their efforts, presumably, are adequately rewarded in the value of the predictions they obtain.

The PAC model spells out one further condition: the *test set and the training set have to come from the same data source.* (Here the data source, which we call *D*, should be viewed as a probability distribution.) If the world changes completely after training, then we cannot confidently rely on any regularities detected in the data during training to be useful when applied to data from a later time. On the other hand, if the nature of the data has drifted, but only a little, then one can expect the learned knowledge to be of some use. For example, a system for recognizing bird species trained on data where all the swans are white will remain useful if tested with a very small fraction of black swans, but as that fraction increases the accuracy may degrade.

Archimedes enunciated a counterintuitive property of the mechanical world: *give me a place to stand and I will move the earth.* The corresponding statement in the current setting is *give me a stable data source and I will predict learnable criteria for it.*

To summarize, Probably Approximately Correct learning assumes a data source that is stable in time and acknowledges two technical conditions: The "approximate" acknowledges that errors are inevitable because rare enough cases will be missed during training. The "probably" asserts that the error control guarantee implied by "approximate" is itself only a probabilistic guarantee. No matter how much effort has been put into gathering the data and processing it, there always remains a small chance that the examples found were atypical. Atypical can mean that even some common examples are missed, not just rare ones. If the training set is atypical enough, then the prediction error of the learned classifier can be arbitrarily large. For randomly drawn data, however, the probability of such an atypical training set is small, so that this second source of error is also controlled.

As previously discussed, in the simplest animals, features are likely realized as signals produced by their physical sensors, such as their photo-sensitive cells. With more advanced animals, including humans, relevant features may exist at multiple levels, from light-sensitive cells in the retina at the lowest level to features that detect more complex fragments of images or higher-level abstractions. The target and features can be symbolic names or the concepts they represent. An example of learning at such a higher level would be the target "chair" with features "legs" or "suitable for sitting on." In turn, the features used, such as "legs," could be the targets of earlier phases of learning. Ultimately, one does get down to physical sensor features.

The simplest form of the PAC model applies to beliefs that are either "true or very probably true" or "false or very probably false," and *not* "true with probability 0.63." (In contrast, inside the learned classifier, there may be numerical quantities that express probabilities.) One justification for focusing on this simplest case is that the features and targets comprise the interfaces at which the various learned classifiers interact. If the target of a classifier is a feature of another, so that together they will be chainable, it is best if the two classifiers have the simplest possible interface.

The extent to which these interfaces are more probabilistic in the human brain is open to debate. Humans often make poor probabilistic judgments. Many of the errors in reasoning that behavioral psychologists report are such failures in estimating probabilities.[5] The failure of our intuitions to estimate probabilities with any accuracy may reflect the fact that these quantities are not readily available, or even computed, in the brain.

Robert Rubin, former banking executive and US secretary of the treasury, discusses his career in his book *In an Uncertain World: Tough Choices from Wall Street to Washington.*[6] His career involved decision-making for investments on Wall Street and for economic policy at the Treasury. He writes that "while a great many people accept the concept of probabilistic decision-making and even think of themselves as practitioners, very few have internalized the mind-set. . . . Sound decisions are based on identifying relevant variables and attaching probabilities to each of them." This passage startled me when I read it. Was there any other way of making principled decisions when faced with many uncertainties? Was this way of thinking novel enough to be worth discussing in the early 2000s? Evidently it was, simply because, without training, the probabilities of the various eventualities we consider in our minds are apparently not readily accessible to our thinking.

Robust Models of Computation

How are we to describe a computational phenomenon? Here physics and computer science part company. In physics, one usually describes a phenomenon in terms of mathematical equations that constrain measurable quantities. In computer science, I suggest, phenomena are described as *robust models of computation*.

I have just described Turing computation and PAC learning. I have done this for two reasons. First, they are both needed for defining educability. Second, they are both examples of robust models of computation.

The word "model" has different meanings in different sciences. The same holds for "computational model." Here I will use the phrase *Robust Model of Computation* (RMC) in a particular and demanding sense, and I will argue that such models are useful for defining computational phenomena. The word "robust" is the operative one and used here in a specialized sense that *the meaning of the model should be the same for variants of the model that have the same intent.*

This robustness comes in two types.[7] As an analogy, consider that you are in a foreign country and have purchased a book of cookery recipes written in a foreign language. What reassurance would you find useful regarding the value to you of the book?

The first type of robustness, which I will call Type A, would be a guarantee that all the recipes in the book are translatable into your language, and following those translated recipes will give the same end result as following the originals. This guarantee says nothing about what the recipes are for.

The second type of reassurance, called Type B robustness, will be a guarantee that the recipes are all for varieties of, say, chocolate cake.

Going back to computational models, Type A robustness is exemplified by the Church-Turing Thesis. This asserts that Turing Machines do capture exactly the phenomenon of information processing. This thesis is among the most firmly established principles known to science. The evidence for it, however, is unlike anything in the empirical sciences. The evidence is robustness in the sense we are discussing here: every other model that anyone has devised with the same intent of being realizable and yet capturing the most general sense of information processing has turned out to have *provably* no greater power. (This is analogous to a claim that no natural language has more expressive power for cookery recipes than your language.) In other words, for every other model considered, it has been shown by mathematical proof that whatever that model can do, so can a Turing Machine. The evidence for

Turing computability being the correct notion is this robustness evidence, which is internal to the theory. This evidence is very reassuring. Without it, we would be right to worry that the Turing Machine, defined earlier, is an arbitrary formalism without consequence.

Type B robustness arises from a second sense in which computations can have variants with the same intent. This is when there is a specification of what the computation accomplishes, but there are many ways of achieving that same outcome. For example, computer programs that sort names into alphabetical order all accomplish the same goal or intent, but there are many different algorithms for accomplishing that end.[8] This is exactly like knowing that the result of following any of the recipes in our cookery book will be chocolate cake—at least as long as all forms of chocolate cake are equally acceptable to us.

PAC learning is a model that has robustness of both kinds. First, it is known that, like Turing computation, many variations of the model specification are provably equivalent.[9] Second, as previously explained, the definition includes a specification of what is being accomplished, namely, efficient learning from examples and generalization with controlled error. Each of these two kinds of robustness provides evidence that the model describes a real phenomenon and is more than an arbitrary formalism. A particular computer program, or an individual flowchart made up of some boxes, need not have either kind of robustness. The educability model will need to have robustness to be meaningful. We shall see in chapter 9 that, like PAC learning, educability has robustness of both kinds.

This discussion of Robust Computational Models is intended as more than a philosophical flourish. It explains the foundation of what is being proposed here. Science of every kind seeks to describe phenomena. The question is: What is the source of validity of the description offered? Most of accepted science involves theory and experience, and correspondences between the two. At this very general level, the same holds here. At a deeper level, what is different is that here we look for more specific internal evidence within the theory that there is this robustness. I believe that this kind of evidence will be used in much future science that is concerned with information processing, whether in humans or in machines.

CHAPTER 6

Integrative Learning

In this chapter and the next I will combine some of the seven foundation stones to give a model with certain learning and reasoning capabilities. I believe that evolution discovered this combination long before the arrival of humans, and that the resulting capability is quite general among animals. In chapter 8 we will see that relative to this model, the final step to educability is not such a big step.

Putting Experiences Together

Sea snails can learn from experience and apply their learned knowledge to new situations. We discussed in chapter 4 a case where they learn to avoid water turbulence. Can nonhuman animals take the next step and combine *multiple* pieces of knowledge that they have learned separately when a situation calls for that? The answer is yes. Pavlov's second-order conditioning experiment, already described, is a demonstration of that. Associating a sound with food and separately associating a black square with that sound enabled the dog to associate the black square with food. Animals of all kinds, including Pavlov's dog, can put together learned information.

This putting together offers challenges. In human education, teaching each subject well in a high school does not necessarily mean that the students will be able to integrate the knowledge they have learned in different subjects. Various educational movements have pointed out this problem and have urged that education be more integrative. Having projects that require the use of knowledge from multiple subjects is one way of pursuing this goal. In the current context I am using the phrase *integrative learning* in a more technical sense, but I am seeking to address the same problem. The problem is that while each act of learning concerns one particular aspect of the world, we need to be able to put together what we have learned in diverse different domains. This putting together is intimately related to *reasoning*.

Learning and reasoning have long been identified as critical aspects of human cognition, but their current formalizations have distinct natures. Learning, when defined as PAC learning, has a probabilistic aspect. The guarantees of PAC learning in any instance apply only with high probability, rather than with certainty. In contrast, reasoning, as considered by Aristotle, for example, and formalized in mathematical logic, conventionally has a deterministic meaning because it emphasizes absolute consistency and absolute guarantees of the correctness of what is deduced.

These formalizations of learning and reasoning are therefore contradictory at their core. *Robust Logic* is a proposal I made some years ago for reconciling learning with reasoning and overcoming this contradiction.[1] To do this, Robust Logic reformulates reasoning in terms of the probabilistic interpretation of learning. This will allow errors in reasoning, but it will mitigate this by providing guarantees on the probability of such errors. Error control on the reasoning process, like that discussed earlier on PAC learning, will prevent the system from going too far off the rails while reasoning.

In everyday life we often have need to combine disparate pieces of uncertain knowledge. For example, we might have learned from experience that a certain coat is more effective against rain than another coat. Independently, we may have also learned from experience that a certain source is a reliable one for forecasting what the weather will be in five hours' time. When we decide to put on that coat on seeing that weather forecast, we are combining those two pieces of uncertain knowledge. We do not need to have had examples where we used the same weather forecasting source and the same coat.

We consider this combining operation as a form of reasoning as it enables us to find solutions to problems that may not be obtainable by generalizing directly from experience. What is different here from conventional logic is that the pieces that are combined, having been learned from experience, may each have uncertainty. The new ingredient we will need here is some guarantee that the combining of uncertain knowledge will keep the uncertainty in the derived deduction under tolerable control.

The suggestion is that in the evolutionary history of life on Earth, learning from experience, as described earlier with the sea snail, came first. This learned knowledge was uncertain in that predictive errors were inevitable. Once such learned knowledge became available to an organism, a capability to combine two pieces of learned knowledge came to be advantageous. Robust Logic aims to capture this capability.

Standard logic, which insists on perfect consistency and correctness, is entirely appropriate for precisely defined formal constructs that humans have created, like mathematics and computer programs. Standard logic has indeed been most successful in its treatment of such areas. It aims to describe fail-proof methods of reasoning. Biological organisms, however, face settings where knowledge is uncertain and cope in a different way.

Those who have sought to understand human reasoning via standard logic have often been disappointed.

Human cognition is a near-miraculous phenomenon. We interact with the world in a highly fragmented way, saying or hearing one word at a time and observing the world one visual scene at a time. Our fragments of experience each day are numerous, disorganized, and often not of our choosing or within our control. *How is it possible for humans to put these disjointed experiences together and obtain as coherent an understanding of the world as we do?*

The answer I propose is twofold. First, we need to use appropriate algorithms for achieving certain definable ends. Second, we need a little further help: the world has to be friendly enough to make these algorithms effective. PAC learning was the start of the exploration along these lines. It needed effective learning algorithms to realize the desired capability, but it also needed the world to provide some useful tasks that are learnable and some time-stable sources of experience from which to learn them.

This chapter and the two to follow extend this theme to the broader capability of integrative learning. These chapters describe the capability needed and what is required from the world to make the capability realizable. The treatment needs to be slightly technical because the phenomenon described is extraordinary enough that it demands a substantive explanation.

Augmenting Learning with Reasoning

Robust Logic extends the basic advantages standard logic offers to the more challenging environment that permits uncertainties. In standard logic, we have so-called *soundness* guarantees. These ensure that if one combines two valid statements in certain prescribed ways, then any conclusion drawn will be valid also. For example, if for three propositions A, B, and C it is believed both that "A implies B" and that "B implies C," then one can combine these to soundly conclude that "A implies C." Thus if "Socrates implies Human" and "Human implies Mortal" are both considered to be valid, then "Socrates implies Mortal" can be derived. The validity of the conclusion will follow automatically from the assumed validity of the constituents "A implies B" and "B implies C" and the method used for combining them.

In any system that allows for the combining of uncertain statements, one needs some *analogous* soundness guarantees. Robust Logic provides such guarantees. This combining will consist of chaining together different pieces of knowledge, now each believed to be true with high probability and not necessarily with certainty. The guarantee offered is that in any conclusion so drawn, the probability of error is limited by the probability of error of the constituent pieces that have been chained together. If the constituent pieces were obtained by learning from data, then the error control of the learning process will bound the errors in each constituent piece. This in turn will permit the soundness guarantee of the Robust Logic to bound the error in the conclusion reached when the constituent pieces are combined by chaining.

While the goal is to mimic the advantages that standard logic offers, the realization will need to be different. One crucial difference between the A *implies* B example of standard logic just used and what is needed when the knowledge is learned is that the analogous statement now needs to be A *is approximately the same as* B. The reason is that supervised learning aims to learn a classifier that correctly classifies not only the positive, but also the negative examples. The same classifier will recognize chairs as chairs and nonchairs as nonchairs. This *approximate equivalence* will be denoted by $A \cong B$. In the logical statement "A

implies B," if A is true then so is B, but if A is false then B may or may not be false. "Human implies Mortal" is not equivalent to "Human is the same as Mortal." Many nonhumans are also mortal. Thus "A implies B" is *not* the relationship that supervised learning algorithms provide. (Putting it differently, the outcome of learning is an *approximate if and only if* condition, while A *implies B* is an *if* condition.)

The bottom line here is that the soundness guarantee of Robust Logic will amount to the following. If one has established that "A is approximately the same as B" and that "B is approximately the same as C," then these can be validly combined to conclude that "A is approximately the same as C." This process of combining two such rules will constitute chaining in the sense described in chapter 4.

Further, Robust Logic provides error bounds on the chained statement "A is approximately the same as C" that follow from error bounds on the two constituent statements. Hence error bounds on predictions will be available not only when they derive from the application of a single learned rule, but also from the chaining of multiple rules. I will discuss this more in the section on the "Soundness of Chaining."

I will call a system that can realize Robust Logic, including this chaining capability, an *Integrative Learning System*, or ILS. It will have all the capabilities of supervised learning, but in addition, *it will have the capability of chaining together multiple learned rules.* The Belfast chimpanzees may have been operating in this way—chaining the disparate uncertain knowledge related to climbing along branches and jumping to grab something. How much chaining enhances the functionality of direct learning is an important question, to which I shall return.

For learned information, uncertain as it may be, there are limits to how many pieces are worth chaining since errors compound as the number of pieces chained together increases. For that reason, one suspects that in everyday life we use only short chains of reasoning. On the other hand, as standard logic exemplifies, if we have rules that we regard as universal truths, then chaining an arbitrary number of them *is* valid. Both the axioms of mathematical systems and the laws of physics have such universality. The quantitative sciences work by allowing arbitrarily long chains and the specific predictions so derived are trusted. The trajectories of spacecraft launched decades ago still produce few surprises, although a successful space launch would have involved the chaining of many pieces of knowledge. The hypothesis I am proposing is that this powerful human capability of chaining a large number of universal rules, as we do in science, arises from the co-option of the evolutionarily earlier capability inherent in Robust Logic, of chaining a small number of uncertain rules.

Robust Logic

Robust Logic is a formalism for describing the learning and application of rules. The rules can apply to situations having some complexity and consisting of several parts. Such rules can be chained as just described. The following is a fuller description of Robust Logic.

At any time, there is a set of *rules* stored in memory that have been previously learned. The *Mind's Eye* is an abstract device in which the description of a particular situation can be specified as a *scene*. The contents of a scene correspond to what one senses in the outside world, as well as one's internal thoughts, plans, and predictions. At any instant, the Mind's Eye applies to the scene those rules in memory that are applicable to the scene. This application updates the description of the scene, thereby making deductions or predictions about it. New rules can be learned.

Old rules can be updated in the light of new scenes experienced. The learning is self-supervised, in the sense of chapter 4, with the scenes providing all the examples and labels that are used. Self-supervised learning is central to Robust Logic.

Figure 3 shows an example of a scene. The situation it describes is that of a *toy duck on a big tiger*. To express this information, the scene distinguishes two objects and correctly associates the properties of *toy* and *duck* with one, and *big* and *tiger* with the other. It also indicates the relationship *on* that holds between them.

More generally, the Mind's Eye contains a number of *tokens* initially undifferentiated, to be used for representing different objects in particular scenes. These tokens are placeholders for arbitrary objects. At any time, there will be a fixed set of *attributes* that are available for describing the objects in a scene. The set of attributes illustrated includes the five attributes big, tiger, toy, duck, and on. A *scene* specifies a particular situation, seen or imagined, in terms of the set of allowed attributes. The scene can be described by a diagram, such as figure 3, or *equivalently* as a set of *assertions*.[2]

Now consider the more involved scene from the world of the crow shown in figure 4. This uses a different attribute set and describes a situation in which "*a seed is floating on water, which is in a glass that has a water level.*"

A scene describes a *particular* situation. General knowledge that applies to *multiple* possible scenes is captured in the form of *rules*. The following are two examples of possible rules, stated informally:

If a stone is added to a glass of water, then the water level in the glass will rise.

If a seed is floating in a glass of water and the water level rises, then the seed will rise.

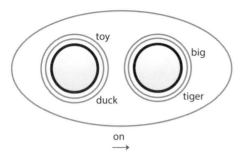

FIGURE 3. The scene shown describes the situation of a toy duck being on a big tiger. This scene contains two tokens, which play the role of placeholders for generic objects. The attributes toy and duck qualify one of the tokens, and the attributes big and tiger qualify the other token. The attribute on qualifies both tokens, in the right order, the arrow indicating which object is on which. The circles with heavily-shaded circumferences represent tokens and the circles with lightly-shaded circumferences and the ellipses, attributes.

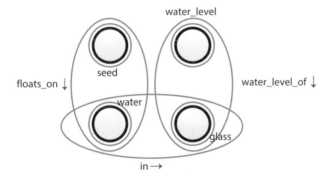

FIGURE 4. An illustration of the scene involving four objects and denoting that a seed is floating on water, which is in a glass that has a water level. From the many scenes experienced, the ILS will learn rules that held for those scenes with high frequency. The rules will be classifiers in the sense discussed in chapter 4.

Since the scene described in figure 4 does not mention a stone, this scene will not invoke the first of these two rules. Similarly, the scene does not mention the water level rising, and the second will not be applicable either. Suppose now that one augments the assertion set in figure 4 with the additional assertion that the attribute stone_added applies to the token representing the glass. To represent this in figure 4, a further circle, labeled stone_added, will need to be added around that token. Then this augmented scene *will* invoke the first rule above, which will then predict that the water level will rise. To express this prediction, the Mind's Eye will *add* to the scene the further assertion rises to apply to the token representing the water level.

The addition of this last assertion, that the water level rises, makes the second rule applicable for the first time. Applying this rule will result in the further update to the scene that adds the assertion that the seed will rise, too. The application of the two rules in sequence is an instance of chaining.

So far, the description of the rules has been informal. We shall need to be more precise later about what is and what is not allowed as a rule. For now, the point is that at any instant some scene is specified in the Mind's Eye, the rules that are applicable to it will be applied to it, and the scene will be updated accordingly. As will become clear later, a rule here is simply a learned classifier in the sense discussed in chapter 4.

Knowledge as Rules

Actions are performed on both scenes and rules. The basic action performed on a scene in the Mind's Eye is to apply a rule to the scene and to update the scene in accordance with what the rule predicts. The basic action on a rule is to update it in accordance with the learning algorithm, using as examples the scenes encountered in the Mind's Eye.

To describe the rules in more detail, we return to the earlier example of the clever crow. In its Mind's Eye, the crow may be imagining the situation that it sees in front of it in the real world, such as a glass of water with a seed floating on it. The crow is contemplating whether to drop a stone into the water. In its Mind's Eye, it will work out how the current situation would most likely develop if a stone were added to the water. It will do this evaluation by applying the rules that are in its memory, which comprise its theory of the world. It will update the scene in its Mind's Eye, the locus of its imagination, as dictated by the rules. The updated scene will describe its prediction about the likely outcome of adding the stone to a jar of water with a seed floating in it.

This, of course, is only a playful illustration. Determining which rules an actual crow would apply is challenging. It is possible that through its evolutionary inheritance or previous learning, the crow will drop an object into water as a reflex response to the situation described, without in any way imagining the outcome or using chaining. The point remains that running the world forward in a Mind's Eye is a useful general skill, for both humans and crows, for situations for which there is no reflex response available.[3]

Let us return to the scene shown in figure 4. In the absence of any other assertions, the scene will express the totality of the information about the situation the crow is analyzing that is available in its Mind's Eye. The ILS has a number of rules, hopefully a very large number, of which only a few will typically apply to any one scene. Those rules that do will be applied to derive the consequences that they imply about the scene.

The following was the rule given earlier: *if a stone is added to a glass of water, then the water level in the glass will rise.* This is an informal statement. Can one say more precisely what is and what is not allowed as a rule?

Since the ILS learns rules by the specific learning algorithm it has, the class of rules allowed has to be learnable. To illustrate the notation used, consider the following example rule:

ForAllTokens [Condition1 OR Condition2 ≅ rises].

The meaning of this rule is that for each token, the attribute rises will be predicted to hold for that token if at least one of Condition1 or Condition2 holds in the scene for that same token. Necessarily, Condition1 and Condition2 will involve some attributes on tokens since otherwise no prediction can be reasonably made specific to that token.[4]

As far as terminology, "Condition1 OR Condition2" will be called the *left-hand side* of the rule and "rises" the *right-hand side*. The approximate equivalence ≅ symbol separates the two.

An example of a Condition1 is one that expresses the condition that *a stone is added to a glass of water*. The rule above with the given Condition1 formalizes the intent "*if a stone is dropped into a glass of water, then the water level in the glass will rise.*" The rule allows for a second alternative, Condition2, which would also be sufficient to predict rises. This condition could say, for example, that some more water is poured into the glass.

Recall that the ≅ sign denotes "approximately the same" in the sense previously discussed. The intent of the rule is to predict whether something will rise. The left-hand side, everything left of the ≅ sign, will then need to account for *all* conditions that would make something rise. This example allowed for two conditions, Condition1 and Condition2.

An ILS would learn a rule that predicts rises in the same way that a generic learning algorithm would learn to predict which images contain a chair. Learning would be from examples of scenes where, for various tokens, rises holds and for others it does not.

The ForAll and ThereExists are standard notions in mathematics and mathematical logic. The statement "for all pairs of different even numbers there is an odd number between them" has an unambiguous mathematical meaning in the domain of whole numbers. However, when describing rules that are of an empirical or common sense nature, it is not so clear over what domains the claims ForAll or ThereExists are made. Our use of ForAll and ThereExists here has no such ambiguity. Here, a claim ForAllTokens or ThereExistsToken refers *only to the tokens in the one scene at hand and to the assertions about them included in the scene description.* This means that, for a given scene, the truth of the left-hand side of a rule can be determined without any need to go into the rest of the world outside the scene. To decide whether the *left-hand side* of a rule applies to a particular scene, one needs only to do a routine check through the list of assertions that specifies the scene. These checks, however, do need to be made for all possible assignments of tokens. Thus, to check "ThereExistsToken water," one needs to see whether in the scene there is *any* token for which the attribute water is among the listed assertions. In the scene described in figure 4, that assertion holds for just one of the tokens.[5]

After applying a rule to get a revised or augmented scene, this process of finding further rules to apply and applying them can continue. These later applications may be dependent on the assertions added by the earlier applications, such as that the water level rises. In the current example this latter assertion will trigger a second rule that then predicts that rises also holds for the seed.

Tolerance of Inconsistency

To recap, the rules learned fit the agent's experience in that they held frequently for past scenes in its Mind's Eye. This implies that these beliefs are approximately consistent with each other on natural scenes that arise in this agent's Mind's Eye. If, for a given situation, one way of chaining rules leads to the conclusion that a certain attribute holds for a certain token, but another chain leads to the opposite conclusion, then we have an inconsistency. This can be shown to occur rarely if all the rules have been learned with small enough error from the same data source.

This gentle tolerance of inconsistency is noteworthy, since any kind of inconsistency poses a formidable obstacle in classical logic. The so-called *Nixon diamond*, formulated by researchers in artificial intelligence in the 1980s, articulates this conundrum.[6] Suppose we have the two logical implications that "Quakers are pacifist" and "Republicans are not pacifist." What happens if we apply these two rules to the person of Richard Nixon, who was both a Quaker and a Republican? If we interpret the rules as absolute truths about the world, then contradictory conclusions would follow, which classical logic does not allow. In Robust Logic, we have only approximate equivalences. Different chains of rules may give different predictions. However, if Nixon and his characteristics occur frequently in the examples seen, then at least one of these two rules will be frequently inaccurate. Learning to higher accuracy is the way to reduce the probability of inconsistencies.

This tolerance for inconsistency has some psychological plausibility. As humans, we can never be certain that a new situation will not arise that shows our firmly held beliefs to be in conflict. Sometimes we know that our beliefs are in conflict, and we choose to live with that contradiction.

Also, as will be discussed later, an individual can have familiarity with several belief systems, such as different political theories, that are in great conflict. A person can entertain and argue about all of them without being committed to or believing in any one of them. In that case, the different belief systems would have rules with different *contexts* (as defined more precisely later), and rules from different contexts would not be chained.

Rules as Learned Classifiers

We can now be a little more specific about the nature of the rules allowed. First, the left-hand side being the OR of several possibilities, as in the earlier example, is instructive since many concepts are defined in terms of a finite number of alternatives. A species of plant has a finite number of identified subspecies, and a word has a finite number of distinct meanings listed in a dictionary. The latter example is particularly relevant since one setting to consider is that of a list of words, from, say, an English dictionary, each serving as the right-hand side of a rule, with the left-hand side amounting to a definition of that word in terms of the other words. If the right-hand side is rises, for example, then the left-hand side will be the definition of rises in terms of other words. If words have a small number of distinct meanings, then it is appropriate to have the left-hand side expressed as the OR of these meanings.

However, the left-hand side will *not* be restricted to such ORs. Much more general forms will also be allowed. *The requirement on the left-hand sides is simply that the resulting rules belong to a class of classifiers that is learnable in practice.* Identifying what these learnable classes are is the subject matter of the field of machine learning. Theoretical results show that some simple classes of classifiers are provably learnable in the Probably Approximately Correct sense. One such class *is* the OR of several variables.[7] There are also some widely used methods, such as deep learning, further discussed in chapter 11, that are used *heuristically* in the sense that one expects them to succeed only on some datasets, but we cannot predict which without trying. Failure on some datasets is not necessarily dangerous. By testing the classifier on data not used during training, one can catch such failures before they are deployed to make predictions on new examples.

When building a computer system that has the capabilities of an Integrated Learning System, we would allow on the left-hand side rules that are as expressive as our machine learning algorithms can handle given our resource budget for data and computation. An OR of several possibilities is an illustrative example. The class of rules our brains use is yet to be determined. The one assurance we can safely rely on, I think, is that this class is a common inheritance to all humans.

Any single rule may be lengthy and complicated. *The rules may need to be complicated if the world is complicated.* Having such lengthy, complicated, and even hard to interpret rules is acceptable if the rules are learned. The rules need to be rich enough to express the complexities of the world, but not so rich that they are no longer learnable. In the practice of machine learning, the learned rules may contain millions of parameters and more. In the human brain, we have on the order of a quadrillion (1,000,000,000,000,000) synapses. There is plenty of reason to believe that learning algorithms, whether deployed in brains or in computers, can be powerful enough to find effective classifiers for real world problems.

One idea adapted here from logic is that of subdividing the example (say, the picture of a chair) into separate parts or sub-objects (different legs represented by different tokens) and expressing relationships among them explicitly. In contrast, the basic supervised learning model treats the input, composed as it may be of many features, as a unitary object.

In colloquial usage, as well as in artificial intelligence research, the word "rule" sometimes suggests that a person has
explicitly constructed the rule. In the current context we assume
otherwise, that the classifiers are learned from data. This allows
rules that are complex and hard to interpret when examined.
What we expect of a rule in this chapter is that it is learnable
from examples and can make appropriately accurate predictions
in a world where situations may be complex and have many
distinguishable subparts. (In chapter 8 we will additionally
allow rules constructed by humans.)

One elaboration of the system is to allow each rule to have a
context. A simple such context would assert that for certain tokens, a certain combination of attributes holds. The rule would
need to hold only when those attributes hold for those tokens.
For example, a context could insist that in the scene there be a
token that represents both French and cookery. This would be
realized by inserting the condition "French AND cookery:"
after the quantifications, such as "ForAllTokens," and before the
body of the rule "[...]":

ForAllTokens French AND cookery: [...].

The rule specified in the brackets [...] would need to hold
only on that subdomain of scenes for which there is a relevant
token that has the characteristics of both French and cookery.

The Power of Learning

We have given two examples of rules that refer to the crow
world. The range of problems that a real crow needs to solve
would require many more rules. Is there some naïveté in thinking that such a system, even with many and more complicated

rules, can express all the knowledge that a crow would need to cope in its complex world?

Several questions arise. Do the rules have enough *coverage* that they apply to a useful range of circumstances? Are the rules *accurate and consistent enough* that they can be usefully applied?

In the early history of artificial intelligence, especially in the work of John McCarthy, there was a central proposal to capture knowledge as a set of *logical* rules, as in classical logic. Humans would compose the rules and the computer would reason using them. This endeavor, of programming a set of rules from which logical deductions would be useful in broad realms of everyday human life, did not succeed. In what way is the present proposal different? The brief answer is *learning*. I believe that learning can deliver the needed combination of the three necessities of coverage, accuracy, and approximate consistency that the logical approach had failed to achieve.

One clue about the awesome power of learning in the sense we have been interpreting it here is provided by the success of machine learning as currently realized by computers. I elaborate on this in chapter 11. A second clue is from biology. In evolutionary terms, survival of an individual organism can be interpreted as a label that the environment provides as a reward for beneficial behaviors. Under that interpretation, Darwinian evolution can be formulated as multiple phases of a weak form of PAC learning.[8] The richness of the inventions of evolution, from the human eye to the flight of birds, each an adaptation to improve the chance of survival, can then be also viewed as illustrations of the power of learning.

There are also clues that are much closer to the discussions already given here. A technical reason for believing that learning can deliver the three necessities listed earlier of coverage, accuracy, and approximate consistency is that the definition of

PAC learning implies all of them. First, the definition promises low prediction error for *all* inputs from the data source. Good *coverage* follows because one measures error over all possible inputs as they occur with their different probabilities in the data source. *Accuracy* follows simply from the promise that the errors can be made arbitrarily small with enough effort. Finally, *approximate consistency* follows from accuracy, since if contradictory predictions occur frequently via different chains of reasoning, then some of the rules must have low accuracy.

There is substantial subtlety hidden in these claims. With the task of having a computer classify images from the web according to whether they contain a chair, there is little mystery. The data source D is the set of all currently existing pages on the worldwide web, which has some stability from day to day. Having predictive accuracy for this stable data source is clear enough as a requirement. The practical effectiveness of machine learning algorithms for image labeling attests to the validity of this approach when there is a stable source of examples: they deliver coverage, accuracy, and approximate consistency for the basic supervised learning task.

The claim that we as humans can exploit an analogous common data source to get the coverage, accuracy, and approximate consistency for all our knowledge is more mysterious since it is more difficult to identify a single source for all the knowledge and beliefs that a single human or the human population has. For images on the web, there is a recipe for accessing the single data source. One can seek to choose one webpage, choosing from among all the existing webpages with equal probability. As humans, however, we have disparate experiences and a variety of belief systems. *In what sense is there a common data source for all the varied experiences of an individual or the worldwide human population?* Chapter 7 will address this riddle.

Chaining with Equivalences

Let us assume for now that we do have a stable data source D. Is the chaining of learned equivalences as powerful as the chaining of implications that is familiar in traditional logic?

The answer is yes—the chaining of logical implications *can* be mimicked by the chaining of learned equivalences. To mimic the implication $A \to B$, we will need to learn an equivalence $X \cong B$ for some X that may be much broader than A. In other words, while for an implication it is sufficient to name *one* condition A that implies B, for an equivalence one needs to encompass *all* conditions that imply B. This may sound burdensome, but it is exactly what supervised learning provides and is good at. If we train a classifier to distinguish between images that depict elephants and those that do not, this classifier will have to recognize elephants of every kind that are common enough in the given data source. The approximate equivalence that needs to be learned could be

$$A \ \text{OR} \ G \ \text{OR} \ H \ \cong \ \text{elephant}$$

where A, G, and H refer to African savanna, African forest, and Asian, if between them these describe all species of elephant. Now the Mind's Eye could use this equivalence for reasoning in the same way as a logician would use the implication $A \to$ elephant. If a scene satisfies A, then it satisfies the left-hand side A OR G OR H of this rule. The rule predicts that elephant is also true for that scene, with high probability. Suppose that we also learn a second approximate equivalence

$$\text{elephant} \ \text{OR} \ J \ \text{OR} \ K \ \cong \ C$$

where J, K might be other animal species, for example. Then for a scene for which A is true, we can apply the first rule to deduce that A OR G OR H and hence also elephant is true. As a second step, we can invoke the second rule, elephant OR J OR $K \cong C$, to deduce that if elephant is true, then elephant OR J OR K and hence also C must be true (all with high probability). In this way, the chaining of these approximate equivalences will mimic the chaining of the logical implications $A \rightarrow$ elephant and elephant $\rightarrow C$.

Teaching by Example

I believe that the cognitive systems of animals originally evolved to operate in a world that, like the PAC model, is learning neutral: the world was not seeking to help the learner learn, nor to obstruct its learning. There are examples of deceit in the natural world, but these usually mislead about the nature of an individual example and are not attempts to misdirect the learning process altogether. Camouflage is an example where an individual provides misleading information about itself. In human fraud, the features of an item for sale may be misrepresented.

On the other hand, political propaganda and public relations efforts can cross the line to interfere with the learning process itself and become *learning adversarial*. By presenting false examples systematically, they can seek to change the classifier the human learns, and as a result the criterion the learner will use to judge future individuals or events. Instead of falsifying the description of an individual or event as the end, the goal is to change the victim's classifier, and thereby judgment, even in situations that are otherwise accurately presented.

In the opposite direction, teaching has a *learning enhancing* role. It plays a crucial role in the enhancement of Robust Logic to educability as described in chapter 8, allowing the transfer of explicitly described rules. However, teaching already has a positive role in Integrative Learning Systems, where there is no transfer of explicit rules. This is what I want to discuss next. What are the ways in which a teacher can help if the learner can acquire beliefs only by learning from experience and not by being taught an explicitly described rule by instruction?

A teacher can be helpful in several ways. One is to provide examples that are difficult to obtain. Suppose the learner wants to prepare for a journey to a distant country to observe rare butterflies specific to that area. If the teacher can provide examples according to a distribution that well approximates the distribution found in that country, then the learner will be well prepared. The learner will still be PAC learning from the examples seen in the Mind's Eye, but the teacher has changed the distribution for the would-be traveler to the one available in the distant country.

A teacher's choice of categories even for the same examples may greatly influence what a child learns, as has been mentioned already. Also, the teacher may be judicious in the choice of training set. For natural concepts, just a few features may be enough to determine the concept. Picture books for young children often have simplified illustrations with only the essential characteristics of the illustrated objects shown. An elephant is just a gray blob with a trunk and two tusks. This reduces the possibility of false associations with distracting irrelevant features and thus reduces the number of examples needed.

If a teacher has many concepts to teach, the order of teaching them may be critical. For two related concepts A and B, it may be that A is easier to learn if B can be used as a feature for A, or even that A cannot be learned without first understanding B. The teacher can drive the order of the concepts by supplying the examples in that order. It has been said that the most important task of an instructor is to choose what to teach. The current paragraph refines this to emphasize the importance of sequencing.

Innate and Acquired Attributes

For humans at birth and machines at manufacture, some knowledge is wired in, as are also some biases in what will be easier to learn later. Life forms can acquire only limited amounts of knowledge during their limited life spans. The rest that is needed will have to be available at birth, provided through millions of years of evolution.

The attributes of the ILS would include features that correspond in biology to sensory information, about sight, sound, smell, touch, and taste. They would also include features that correspond to *inner* experiences, of pleasure, pain, empathy, and hunger, among others. These would all give rise to attributes partially set in place by evolution. We also have attributes that are acquired mostly from experience. Some attributes could be set up first as targets for memorizing *instances*, as discussed at the end of chapter 4. For example, the attribute corresponding to the appearance of a particular caregiver would be so acquired early in life. Attributes corresponding to physical landmarks are also examples where memorization of instances is useful.

Imagination and Novelty

The question now is whether the mechanisms of integrative learning, as described, are sufficient to give rise to phenomena we conventionally associate with imagination and novelty. A position on these questions is necessary since these qualities have been essential in the emergence of civilization.

The integrative learning framework has a front-and-center place for imagination. It is the Mind's Eye. It is there that the ideas being entertained evolve from those entertained an instant earlier. The mechanism of this entertaining is the application of the rules, sometimes influenced by current sensory input. The rules have a variety of natures. Some may be inborn and supported by the experience of our ancestors as captured by evolution. Others are learned through individual life experiences. Others still will have been taught by instruction and may have arbitrary relationships with any realities in the world. The content of our Mind's Eye at any instant is determined both by the external stimuli that we experience at the time and by the rules we have in our memory, which incorporate our past experiences both evolutionary and direct.

In any one physical situation, different people may fill their Mind's Eyes with different content and may be interpreting the same external situation differently. When considering eventualities beyond our immediate senses, such as what is likely happening in another place, or what will happen at another time, our imaginations can run wild along tracks strongly biased by our previous experiences.

Being able to run forward various possible sequences of events in our Mind's Eye may help us make better plans and decisions. It may also help us isolate what extra knowledge would be most useful to us in our planning and decision-making and so enable us to ask good questions, which is an important human faculty.

What is the source of novelty? How is a concept invented for the first time in someone's mind? How did someone first conceive of an ellipse or a logarithm?

As we defined it so far, at any *one* time there is a fixed set of attributes. These are the only candidates for being the right-hand sides of rules, the attributes predicted, or attributes in the left-hand sides of rules, on which the prediction depends. How can we extend an ILS to allow *novel* attributes to be created for the right-hand sides and in the left-hand sides? The answer is to allow *combinations* of existing attributes to be added to the attribute set. A simple method of combination is an AND. For example, if the current fixed set of attributes includes both French and cookery, then French_cookery will be such a combined single new attribute. Allowing rules with these new combinations treated as *single* features on the left-hand sides and as targets on the right-hand sides, will permit an ever-growing set of attributes.[9]

An ILS can acquire novel combinations in several ways. Perhaps the more mundane way is to internalize combinations of attributes that the *environment suggests* are significant. Certain phenomena in the external world are very distinctive—a hurricane is difficult to miss the first time you experience it. When a hurricane is experienced, a set of already recognized attributes is made true, such as strong_winds and warm_ temperature. This combination can be retained as a new compound attribute that will be useful in thinking about hurricanes in the future. The environment can also suggest significant information in the less direct way of having two or more existing attributes co-occurring more often than they would by chance. Instances of the concept "animal" have features that they usually share, such as being alive and being able to move. An individual might be able to detect such highly correlated sets of features and use their co-occurrence as the way of discovering that there is an important concept corresponding to "animal." These give two ways in which the environment can suggest new attributes to us without the mechanism of a teacher or supervisor. One works from the unexpected nature of an individual example; the other, from the statistics of many examples. Both can be viewed as unsupervised processes.

An equally basic question is how a useful new combination can be *internally generated* in the mind. If you are exceptionally tall and want to buy a car, you may invent the concept of "car with much legroom," where the two components, "legroom" and "car," are each already familiar to you, but the combination is not a standard one you had been told about. It is the learner who decides here that a particular new *compound attribute* will be of use. The freedom to choose which compound attributes to consider is one of our great freedoms.[10]

There may be no time for considering all the possible combinations of all the concepts we recognize. Some combinations will turn out to be more useful to an individual than others. It is often said that the most crucial step in scientific research is that of formulating the right question to ask. New questions often involve making a distinction that splits into two what was previously thought to be a single concept. Perhaps a new qualification is imposed to subdivide an existing concept and so form a novel conjunction. Gravity + Attraction = Gravitational Attraction, Magnetism + Field = Magnetic Field, Selective Advantage, Algorithmic Complexity, Squashed Circle are all examples. I do not believe that there is a universal method for choosing the best new distinction or conjunction to make. Different individuals may be using different strategies for considering such new combinations. The effectiveness of their choices will vary.

Novelty rarely arises in the form of something entirely without precedent. It is usually a new combination of the old. The educability framework encompasses this kind of novelty at its core.

CHAPTER 7

What Integrative Learning Does

Compounding Knowledge

Suppose you want to study two topics, A and B, say physics and mathematics, or history and literature, by taking a separate course in each area, but you want to be able to integrate what you have learned. You will need at least two things: First, there should be some subject matter to which you can apply what you have learned in both courses. Second, the content you have learned in the two courses should be combinable so that the conclusions you reach are valid.

Life is much more challenging and disorganized than taking two courses. Every moment we can be faced with a new situation and receive new information. How do we integrate what we learn from such disparate experiences? How do we know which knowledge is applicable in which situation?

Life-forms have had to face these challenges from the beginning, and evolution needed to address them long before humans arrived. Reflecting this, these challenges are addressed already by Integrative Learning Systems—we do not need to wait until the chapter that follows where we finally get to educability.

An ILS learns general rules from individual situations represented in its Mind's Eye and applies these in combination to new situations. This capability generalizes PAC learning and in doing so introduces two sources of complication that formalize the two challenges just spelled out.

The first complication arises from the combining operation. Even if each of the learned rules is accurate when applied alone, how do we know that chaining them will maintain accuracy?

The second complication is that, unlike the idealized supervised learning formulation, maybe no one is feeding examples from a fixed carefully curated data source. How can we conceptualize the commonality that an individual's experiences at different moments share? I will address this second puzzle later using the metaphor of the Patchwork Tapestry. First, I consider the validity of chaining.

Soundness of Chaining

As previously discussed, in classical logic $A \rightarrow B$ denotes the *exact implication* that "if A is true then B is true." It is axiomatic that two rules $A \rightarrow B$ and $B \rightarrow C$ can be validly chained together to conclude $A \rightarrow C$. In the semantics of PAC learning, we desire the same, but now in a probabilistic sense. As already noted, it is not exact implications that are being learned here, but *approximate equivalences* $A \cong B$. The main difference is that in the former case of an implication $A \rightarrow B$, if A is false then nothing is implied about B, which may then be true or false. If we had exact equivalence, then either both A and B are true or both A and B are false. With the approximate equivalence $\cong$ that we have here, the equivalence is true only with high probability.

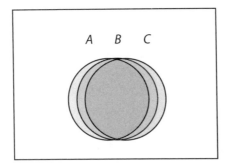

FIGURE 5. The interior of the rectangle represents all possible examples in the data source. The insides of the circles (*left to right*) represent the examples on which the assertions *A*, *B*, and *C* are true, respectively. One throws a dart at this rectangle and considers the probability of hitting within each of these circles. If the probability of hitting within either *A* or *B* but not within their overlap is at most 1 percent and the same holds for *B* and *C*, then it follows that the probability of hitting within either *A* or *C* but not within their overlap is at most 2 percent. This is because any point in which *A*, *C* differ must be in one of the crescents in which *A*, *B* or *B*, *C* differ. This is true with no preconditions. For example, the different parts of the rectangle do not need the same probability of being hit.

Thus $A \cong B$ means that for examples taken from the data source *D*, with high probability *A* and *B* will be both true or both false. With small probability, one will be true and the other false.

We want to be able to chain the approximate equivalences $\cong$ just as we chain the certain implications $\rightarrow$. The meaning is that if $A \cong B$ holds with probability at least 99 percent on the natural distribution *D*, and if $B \cong C$ also holds with probability at least 99 percent on the same *D*, then $A \cong C$ will also hold on this same *D* with *some* minimum probability. In this instance, as illustrated in figure 5, a probability of at least 98 percent is implied. In this sense, therefore, we can validly chain the pair of equivalences $A \cong B$ and $B \cong C$ and deduce $A \cong C$, though with increased error tolerance.

World without Reason?

Despite the self-evident usefulness of chaining, there is a potential paradox lurking in the background. Consider the case above that $A \cong B$ and $B \cong C$ both hold and are learned as two separate rules. Then on the presentation of a new scene for which A holds, the two rules will be chained to deduce that C also holds, as if $A \cong C$ had been learned directly from the data.

The potential paradox is that if it is the case that both $A \cong B$ and $B \cong C$ are learnable from examples, why could not $A \cong C$ also have been learned? If this approximate equivalence $A \cong C$ held on the training data, why was it not learned? Why was there a need for learning the two rules separately and then chaining them? Why is the result $A \cong C$ of the chaining not as easily learned from the data directly as are its constituents $A \cong B$ and $B \cong C$? Perhaps, wherever learning is feasible, chaining is redundant because one can obtain the same result as chaining by learning the more complex single rule and applying it directly.

This question of redundancy arises for any reasoning system, including classical logic. The result of an act of logical reasoning is, by definition, already implicit in the information that is available at the starting point. Reasoning in classical logic expands knowledge only tautologically. The argument for reasoning not being redundant needs to be of a quantitative nature. For example, in classical logic deducing a proposition of a given complexity from some others that logically imply it, is quantitatively useful to anyone who does not have the resources to precompute and store all the implications of what they know up to that complexity. For Robust Logic, some of the nonredundancy arguments will also be quantitative and for a similar reason. Some such arguments will explain, for example, that learning $A \cong C$ directly is *harder* than learning $A \cong B$ and $B \cong C$ separately and not that it is impossible.

The following are two such quantitative arguments for why chaining is *not* redundant.

First, during learning one would generally set a maximum error level, say 1 percent, above which a rule would not be considered reliable. Rules that the training data does not support to that accuracy would not be learned or used for chaining. A rule that the data supports at only the 2 percent error level would be missed. On the other hand, we will have to tolerate deductions that have a higher probability of error. This was the case in the example given of $A \cong B$ and $B \cong C$ both holding with probability 99 percent, and the deduction $A \cong C$ then being guaranteed only at the 98 percent level. We always learn to one level of accuracy and then use chaining to obtain implications that may hold at a less stringent level. We could learn all the rules that hold at the less stringent level but may not wish to because there may be too many of them to learn and store.

Second, instead of the simplest pair of rules $A \cong B$ and $B \cong C$, consider a more complex pair of the form $(A \text{ OR } E) \cong B$ and $(B \text{ OR } F) \cong C$. Here the effective result of chaining is $(A \text{ OR } E \text{ OR } F) \cong C$ since by the first equivalence we can replace B by $A \text{ OR } E$ in the second equivalence. Then one case of nonredundancy is if the left-hand sides $(A \text{ OR } E)$ and $(B \text{ OR } F)$ are separately simple enough that they fall within the learnable class, but the chained rule, $(A \text{ OR } E \text{ OR } F) \cong C$, is outside the class and therefore not learnable. For example, E and F may themselves be the ORs of ten distinct attributes each so that $(A \text{ OR } E)$ and $(B \text{ OR } F)$ could be each written with eleven attributes, but the compound concept $(A \text{ OR } E \text{ OR } F)$ would need twenty-one.

There is a third rationale for the value of chaining, which we discuss next.

Modularity and Partial Information

The world has much *modularity,* which means that it is composed of many distinct parts that are largely independent of each other. Knowledge about seeds floating on water and knowledge about different elephant types do not have much commonality. This modularity, a helpful property of the world, allows our working memory, or the scenes in a Mind's Eye, to contain at any instant only *partial information* about a real situation and still function. It only needs the information that is relevant to the decision at hand.

When training a conventional machine learning system to distinguish images that contain chairs from those that do not, each example contains the values of all the features that will enter the decision. In other words, each training example may be a $1,000 \times 1,000$ array of pixels and each test example will be of the same form. The learned classifier will be determined by the values of the millions of pixels of the training examples and the provided label of each example. When making a prediction, the information used about the test example will be *its* million pixels, which correspond one-to-one with the features of the training examples. We call this learning from *full information.*

This is different from the example we gave earlier of learning that a certain coat is more effective against rain than another and learning separately that a certain source is more reliable for forecasting the weather than another. In this case, the examples needed for these two acts of learning will have disjoint features. In one, the examples are experiences of wearing different coats in the rain, and in the other, memories of weather forecasts and the ensuing weather. The features that matter in the two cases are different. The examples seen in the two cases have somewhat disjoint feature sets defined.

One commonality among these limited partial views of the world, such as of rainwear and weather forecasts, is that the agent will be viewing both through their one Mind's Eye. Each example will have the form of a scene that asserts some attributes on tokens. The various scenes refer to distinct situations in the world. They will make assertions about different aspects of the world and be silent on multitudes of other aspects. One scene may be about three coats next to each other, another of someone walking and getting wet, and a third about the weather forecast on a cell phone. What is the common distribution or data source D that encompasses all this?

The answer is that one can consider a single large set of attributes in terms of which *all possible scenes in the Mind's Eye with the given set of attributes can be described*. For example, a first attribute could represent the assertion coat, a second weather_ forecast, and a third wet. We would also have compound attributes as described in the previous chapter.

The set of compound attributes that can be constructed from the vocabulary of attributes and tokens serves as a universal set for everything expressible in this Mind's Eye at that instant. In this world of partial information, however, in any one scene the values of the vast majority of these attributes will not be determined in the scene. To permit this, we need to allow that an attribute, besides having the possible values of 1 and 0 for true and false, respectively, can also have a third value of $, which we shall call the *obscured* value, to indicate that it is hidden from us.[1] This is what we call partial information.

In this context of learning from scenes, any of the attributes can serve as a target as well as a feature. Unlike the standard formulation of supervised learning, there will not be just a single target of learning. The role of learning, then, is that of *uncovering relationships among the attributes from example scenes in which they are not obscured and predicting them in scenes where they are obscured.* Rules involving both stone_added and rises would be learned from examples in which they both occur, while rises would be predicted in an example for a token for which the value of rises is obscured.

Was Pavlov's Dog Right?

Recall Pavlov's second-order conditioning experiment. In one context, the dog learned to associate a sound with food and there was no black square present. In another context, it associated the black square with the sound and there was no food present. It then chained these two associations to realize an association between a black square and the presence of food. Living entities may realize chaining in a variety of ways. Some pairs of rules may be chained to become hardwired in the neural system as one rule. Chaining realized in an ILS offers more flexibility than this in the ability to combine a new set of rules as demanded by each new circumstance.

The question I want to ask is: Was the dog justified in associating the black square with the coming of food? The answer is: *It depends.* It depends on the world it lives in. In the real world, where food does not often follow black squares, it may have been wrong. In the experiment the dog appears to be making the *independence assumption.* In this world there is much modularity. Therefore things learned in different contexts often hold independently of one another and can be chained together. This justifies the dog adopting such an independence assumption as a default policy. Nonetheless, this policy may result in unfortunate outcomes in situations where independence fails badly.

Exploiting Modularity

Suppose I want to take my dog from Boston to New York. I have never seen dogs taken from Boston to New York and have zero examples of this having been done. But I know two general rules that are supported by examples I have witnessed. The first says that dogs generally can be put into cars. The second says that cars generally can be driven from Boston to New York. Neither rule is universally true. Some dogs are too large for some cars, and some cars are not in a good enough condition for such a long journey. The risk I am taking in chaining these two rules—and therefore deciding to try to drive my dog to New York—is that of assuming that there is not a hidden dependence between their failure modes. In principle, there could be a New York state law that forbids dogs being driven into the state.

To formulate the issue more precisely, consider the nature of the information on which we base our two beliefs that dogs fit into cars and cars can go from Boston to New York. For the former we have seen examples of dogs getting into, riding in, and getting out of cars, but in these sightings the origins and destinations of the journeys are not given. For such scenes "dog is in car" would have value 1, but "destination is New York" would have the obscured value, which we denote by the symbol $, as explained previously. Similarly, I may be aware of many car trips from Boston to New York but not know whether a dog traveled in any of them. In this circumstance, I will have learned the two rules from two different datasets with partly disjoint feature sets. From these datasets, I have no information as to whether the feared New York state law banning the importation of dogs is in effect. Chaining will combine them as if the two rules were independent.

The power of chaining is that we can draw conclusions for situations that are rare in the sense that even similar situations are rarely or never encountered. Our experience provides no direct statistical support for the decision we need to make, but it may provide support for several smaller decisions that when chained address that decision. Long chains magnify the role of the independence assumption. At each stage of chaining, we can make the independence assumption anew and be successful in the reasoning if independence happens to hold.

I believe that everyday common sense reasoning uses chaining to get answers for situations that are rare. The rarity of the situation provides the necessity for chaining. For common situations, we can base our decision on a single rule learned from experience. Where the independence assumption holds, we can reach conclusions by learning on smaller datasets and then chaining. Without independence, much larger, possibly impractically larger, datasets would be needed.

The Patchwork Tapestry

I suggest that the way that Robust Logic handles the problem of partially specified examples answers the riddle of how people can assemble effective responses to a complex world they experience as fragmented and disjointed episodes. The answer rests on the assumption that the world is modular, consisting of many different modules or clusters of information that are largely independent of one another. The world we experience is like a *patchwork tapestry*, consisting of many patches. Each patch corresponds to a different information module. Each module is jointly specified by the set of attributes that are defined in it and a context that may be additionally defined in terms of the value of those attributes. Of the former kind are patches that refer to dogs in cars, or cars traveling between cities. Patches of the latter kind could impose a context such as that the location is in New York. These different patches do refer to the same world. In each patch rules can be learned that reliably relate the features or attributes defined in that patch without dependence on attributes that are not defined.

In PAC learning we learn from a fixed data source, say, pictures from the web. Robust Logic rests on the same assumption, but it becomes a little more mysterious what the common data source is if the examples are only partially specified, which they need to be in a complex enough world. We have the worlds of the crow, elephants, raincoats, dogs in cars, car trips between cities, French cookery, and Chinese cookery. These settings are disparate. Yet, as the underlying theory explains, one can identify a common D, which in the metaphor is the whole patchwork tapestry. An individual will specialize to learn well only a few of the patches and will not have the time to learn them all. As mentioned, a patch may be defined in terms of which attributes are not obscured (e.g., everything to do with cars and dogs in them) and, additionally, in terms of the values of these attributes (e.g., whether the Chinese cookery feature is true or false), which could serve as an indicator of context.

Chaining remains sound even in this setting of partial information. For a rule to be approximately correct here, it needs to be approximately correct on the attributes that are explicit in it, irrespective of the values of the obscured variables. This may seem like a tall order on the world, but it is what modularity means. Whenever it holds, learning from moderate-size training sets will be possible, and the soundness of chaining can be exploited.

The world's modularity means that we can learn about the world one patch at a time. If two patches have different identifiable contexts, then their rules may be different, as in Chinese versus French cooking, without making the overall system of beliefs inconsistent. Of course, some belief patches do contradict others. Two political beliefs that claim to refer to the same society but take different views on many situations will be inconsistent enough that an individual cannot commit to both. To make sense of the world, we humans need to make choices among the patches when committing to beliefs. Further, the patches we commit to need to agree at the seams.

CHAPTER 8

The Capability of Being Educable

The Crux

The previous two chapters have described Integrative Learning Systems (ILS) with their capabilities for learning from experience, for scene analysis, and for chaining and applying what has been learned. These capabilities, impressive as they may be, already abound in the animal kingdom. In what directions does such a system need to be augmented to make this system educable? My answer is that *the capabilities of acquiring explicit rules by instruction and of applying these rules need to be added, and the system needs to be enhanced to fully support symbolic names.* Note that the ILS does not require symbolic names, though some animal species that have the ILS capability may also have some capability for symbolic names.

The final definition of an Educable Learning System, or ELS, that I will arrive at will realize the three pillars of educability defined in chapter 4. Recall that the three pillars were (a) learning from experience, (b) being teachable by instruction, and (c) combining and applying theories obtained in both modes. The learning from experience pillar (a) was described in the sections "Learning from Examples" and "Generalization" in chapter 4, and "Probably Approximately Correct Learning" in chapter 5. The ability to chain and apply theories learned via pillar (a) was the subject of chapters 6 and 7 and is an important part of pillar (c). However, pillar (c) incorporates the ability to chain and apply not only rules learned from experience but also rules acquired through instruction, all on top of a Mind's Eye and symbolic names. Explaining these capabilities that extend an ILS to an ELS is the goal of the current chapter.

An Educable Learning System

An *Educable Learning System* includes the ancient ability of an ILS to learn from experience and to apply the learned knowledge in new combinations. But it magnifies its capabilities with a further facility to acquire explicitly described knowledge by instruction. An individual who can be taught a rule by instruction will not then need to pay the price of learning that rule directly from experience. The price of that direct learning might have been high in terms of the danger to which the individual was exposed in gaining the experiences or the time and effort needed.

The hypothesis is that in the course of evolution our ancestors acquired the capability of an ELS. The capability emerged in the genus *Homo*, or a close predecessor. I do not need to make detailed hypotheses about the order in which the constituent components first appeared or came together on Earth.[1] I would guess, though, that rules with symbolic naming taught by instruction were the most recent addition.

Our complex civilization is rich in hard to discover knowledge that requires symbolic names to express. Once one can easily communicate a piece of knowledge to others, many can add to that piece without needing to rediscover it. While the significance of each individual addition by itself may be modest, these additions accumulate to form the vast edifice that is now the total knowledge of humanity.

This suggests the answer to the age-old question of what uniquely human characteristic is responsible for the creation of civilization. Knowledge and culture abound in the nonhuman world too.[2] The issue at hand here is the cognitive capability that makes possible the creation of knowledge and culture on the vast scale that it occurs in humans. My answer is that this capability is educability, *an integration of the abilities to use symbolic names flexibly, to learn from both experience and instruction, and to apply and chain knowledge obtained in both ways.*

Why is this combination so powerful? The ancient capability of learning from experience permits the acquisition of uncertain knowledge. Chaining such uncertain knowledge to any large depth would degrade accuracy sufficiently to become useless. An important part of our civilization, however, concerns knowledge that we treat as holding universally, at least under stated assumptions, as in mathematics or physics. In those cases, chaining to arbitrary depth is valid. Science uses chaining to great depth to obtain some of the more surprising scientific predictions and technological consequences. It is in this context of science and technology that the chaining of universal rules achieves its stunning power. Biological evolution may not have selected directly for this capability. We may be its beneficiaries by serendipity.

The two essential augmentations to the ILS needed for educability are therefore symbolic names and rules acquired by instruction. We shall take them in this order.

More on Symbolic Names

It is easy to make glib statements about human capabilities with symbols. For any explanatory theory of what this capability is and how humans acquired it, one needs to consider how it might be possible to realize it in a brain. This is a challenge by itself. The flexible use of symbolic names apparently took a long time for evolution to discover. Computers may appear better suited to the task, but even there the solution is not so trivial.

Suppose that in a computer program we want to assign to a variable the year that René Descartes was born. We can give that variable an arbitrary name, such as "Descartes' year of birth" or "Socrates" or "xrt452," and then give that variable a value, such as 1596. How does one realize such an assignment between a name and a value in a computer?

An important aspect of digital computers—which, as far as we know, brains lack—is that they have an *addressing mechanism*. Each location in a computer's memory has a numerical address, such as 3,467,236. To assign Descartes' year of birth to memory location 3,467,236 requires two steps. First, we need to associate the sequence of letters in the name "Descartes' year of birth" to that numerical address 3,467,236. Then we need to assign the value 1596 to the location 3,467,236.

If, later, we wish to recall the stored value of "Descartes' year of birth," we need to use that same association, so that "Descartes' year of birth" leads us to address 3,467,236. Then, by reading the contents of that address, we will have the desired answer of 1596.

How do we do the first step of associating the sequence of letters in the name "Descartes' year of birth" to that numerical address 3,467,236? We do not care where in a computer's memory we store a particular item as long as we can find a different place for each item and can keep track of where we have stored each item. We therefore need a practical way of associating an arbitrary name with a distinct numerical address, it being immaterial which address. This is usually done in two stages.

In the first stage, the computer maps the sequence of symbols in the name to some number by a *code*. For example, we may associate each letter of "Descartes' year of birth" with its numerical place in the alphabet, such as, $a = 01$, $b = 02$, ..., $z = 26$, and other codes for other symbols, and string these numbers together as a sequence. Then "Descartes' year of birth" becomes the number $0405 \ldots 08$, which we shall call x. (Here the "D" has become 04 since it is fourth in the alphabet, the "e" has become 05 since it is fifth, etc.)

In the second stage the computer uses a *hash function*[3] that maps the code x obtained in the first stage to a memory location y. With very short names and a large computer, we may get away with choosing location $y = x$ for storing the contents. But this would be wasteful of memory since most locations would not correspond to useful codes and therefore would not be chosen to store anything. Another problem arises if the names are long. Then the x obtained by the encoding could be a number far larger than the address of any memory location. To solve both problems, the hash function will map arbitrarily large numbers x that are produced by the encoding of arbitrarily long names, to a fixed range. If the computer has 3,001 memory locations, for example, then the fixed range of the addresses y could be $0, 1, \ldots, 3,000$.

What properties does a hash function need to have? In general, it is a function that maps the x to a number that is guaranteed to be in the range $0, \ldots 3,000$ in a *pseudorandom* way. Pseudorandom here means that it has some of the properties of pure randomness. The most relevant property of randomness here is that there are few collisions, a collision being a situation in which more than one x is mapped to the same address y. For example, if 100 balls are thrown randomly into 3,001 bins, most will go into separate bins, but some may go into the same bin as others. The latter collisions need to be minimized, since if for a new name the same location is chosen as for a name previously assigned, then we would need to take an extra step to find a previously unallocated address for this new name.

Pure randomization has this good property of keeping collisions low for any sequence of xs. However, for any one code x the address y to which the computer allocates it needs to be derivable from the code, since we *will* want to use that y more than once, typically first to store the value, here 1596, and later to retrieve it as needed. Hence there must be a repeatable deterministic way of generating, whenever needed, the *same* address y from the code x. A true random choice for y each time will not do, since for any one x it would not generate the same y every time.[4]

The paragraphs above discuss *where* we would locate the value of a symbolic name in a digital computer. As we just saw, this choice is not trivial even in a computer. It is much more challenging in a biological brain. The problem is that our brain, with its hundred billion or so neurons, has no addressing scheme comparable to that of a computer, as far as we know. In other words, it does not have a mechanism that, given a number y, accesses a particular neuron, or set of neurons, corresponding to that y.

How do brains do symbolic naming then? In earlier work, this author identified naming as a task performed by the brain that needs explaining.[5] My suggestion there had two main aspects. The first was that what corresponded in the brain to a storage location in a computer was a set of neurons, later called a *randset*.[6] This was consistent with the classical notion of *assemblies* proposed by Hebb in 1949 in the use of sets of neurons.[7] It differed in that, unlike most instantiations of the Hebb theory, it needed no assumption about these sets of neurons, such as that they are better connected to each other than to other neurons. They could be arbitrary sets of a certain size. The second aspect was that what corresponded to the hash function realized in the computer with numbers was an analogous hash function, but now one *realized by the connections* in the brain.[8] The number of connections in the human brain far exceeds the number of bits of information in human DNA. Randomness during the development of a specific brain starting from the embryo is believed to play a role in determining the actual connections. In brains, randomness in the connections will have the effect of making it unlikely that the sets of neurons chosen for different tasks will have abnormal overlap. The xs for which an individual will need to assign neurons are determined by the challenges life has thrown at the individual, much as the sequence of xs in a computer application. The connections in the brain, containing so much randomness, will be ready to deal effectively with whatever is asked of it, just like a computer is ready.

The xs encode the names of the items being stored, whether in computers or brains. For humans, one can view words in a natural language such as English as codes or names. This provides a concrete explanation of the usefulness of language. A primary function of words, then, is to assist in assigning different enough sets of neurons for referring to different items that are stored in the brain. Words in a language are useful for other reasons also, but if they failed in the requirement of facilitating memory, then the rest would not even arise.

Expanding on this a little, the benefit of words and names is to limit the complexity of referring to what they mean. It would be possible to invent a separate name for the concept of

"Descartes' year of birth," though we do not need to since the phrase is short enough. However, if Descartes did not have a name and we had to refer to him as "that son of Jeanne Brochard who became a philosopher," and if there was no word for the concept of "year," then the reference to Descartes' year of birth would be somewhat complex. Having plentiful words and names allows us to refer to the concepts we need as xs of suitably limited length. Here brevity helps.

Once we have localized a symbolic name x, whether in a computer or a brain, we can start performing computations in terms of it. In a computer, the value of the name x would be stored in the address y associated with that name. In a brain, the neurons associated with the name would fire when processing that item. For example, the neurons associated with the word "Socrates" would fire whenever one thinks of that philosopher, or dog.

Physical sensors and actuators are hardwired into a system at manufacture or into an organism at birth. In both cases, subsequent learning from experience may influence the adaptation of this functionality. In living organisms, for example, there are critical periods during which this adaptation is most effective. Physical sensors and actuators, even if subject to such adaptation with experience, are constrained by some such legacy. In contrast, symbolic names suffer from no physical legacy burden. Symbolic names are universal in that they can be anything (though in brains it helps to be not too long) and can refer to anything.

A teacher can provide the learner a new symbolic name as the target for learning a new concept and give examples of it. The concept will denote whatever the teacher is setting out to teach. There is no relationship necessary between the symbolic name and its meaning. The teacher can take any concept, such as "generosity," and call it "x3y." Such a concept can subsequently also appear as a feature that occurs on the left-hand side of rules for other concepts.

Symbolic names are not unique to humans. When a nonhuman animal learns its name, or a word in a human language, as demonstrated by chimpanzees' use of sign language, it is giving evidence of an ability to process symbolic names. An ELS can handle symbolic names in a particularly flexible way, as the features and targets of chainable rules, whether learned or taught by instruction. This flexibility may be unique to humans.

Rules Acquired by Instruction

To recap, in an ILS all rules are learned from examples. In an ELS there is a second method: a rule can be acquired by being given an explicit description of it, for example, as a sentence in English. Rules so acquired are *rules acquired by instruction,* as opposed to *learned rules.* The ELS will be able to apply both kinds of rules to a scene in its Mind's Eye, while an ILS applies only learned rules. An ELS is able to chain together several rules, including a mix of rules of both kinds.

We shall call the class of rules that can be acquired by instruction C^* in analogy with C, the class that can be learned from examples. The general form of a rule acquired by instruction will be like that of a learned rule to the extent that it has a target attribute on the right-hand side that it predicts and has a left-hand side that is a condition on the values of the attributes in the scene at hand.

This gives a second distinct role for human language in educability. The first was to supply symbolic names. The second is to provide a means of transferring an explicit description of a rule to the learner. If a rule involves symbolic names on both the left- and right-hand sides, it is difficult to imagine how such a transfer from one individual to another would be possible without the expressive power of language. For solving a physical puzzle, such as the one discussed earlier about food in a tube, a physical demonstration of what is to be done can be viewed as a method of instruction that does not use language.

One sense in which a rule acquired by instruction can be different from a learned rule is that the connection between the left- and right-hand sides no longer needs to be the approximate equivalence $\cong$, a restriction that was imposed by learning. It can now, in addition, also be an *update* sign $\Rightarrow$, closer to the conventional logical implication. This update $\Rightarrow$ sign in a rule acquired by instruction means that when the left-hand side is made true in a scene, then the Mind's Eye will update the scene to make the right-hand side attribute also true. If the left-hand side does not satisfy the scene, then no update will occur.

The $\Rightarrow$ sign is an operational programming instruction for updating the contents of the Mind's Eye. Learned rules also have a similar operational effect in being able to update the Mind's Eye. The difference is that while the learned rules carry an upfront guarantee that the update is (probably approximately) consistent with the learner's previous experience, the rules acquired by instruction carry no such guarantee. In the latter case the learner is at the teacher's mercy as far as the consequences of acting on the rule.

An ILS operates with learned knowledge that empirically fits the experience of the owner. An ELS is an enhancement that can additionally process rules that are not constrained to fit that personal experience. The features and targets involved in the rules can be arbitrary abstractions divorced from the owner's experience. If these rules acquired by instruction capture deep truths about the universe, then great benefits can follow from applying them. However, dangerous falsehoods can be acquired equally easily through this ability.

Sometimes there are ways of gauging the plausibility of rules acquired by instruction. If the rule relates to the world of experience, then the owner can check its validity against examples. The owner can also judge a rule by its provenance. Has the teacher or book that was the source of the rule provided rules in the past that checked out against experience? Is the source reputable? This issue is the subject of the section "Belief Choice" later in this chapter.

Even if we have ways of evaluating the reliability of single rules, when chaining together a mix of rules, errors may compound. For example, we may easily recall snippets of information we have recently gleaned from the media. Keeping track and correctly combining our confidence level in each snippet may be beyond us. It is possible that we humans have a tendency to apply the rules of our belief system with abandon. For the evolutionarily ancient purpose of learning from direct experience in a stable physical environment, this abandon is justified by the probabilistic guarantees on learned rules and the soundness of chaining them. However, if we add rules acquired by instruction to the mix and apply them with abandon, then all bets are off.

Scene Management

The systems described up to now are not complete. Even the basic ILS needs additional management tools. For example, when is one scene removed from the Mind's Eye, and when is a new scene introduced?[9] Organisms having ILS capabilities need inborn heuristics for managing these tasks and coping with the limited resources that the Mind's Eye, or working memory, has available. Beyond inborn heuristics, can one learn good management strategies from experience? If one is educable, can a teacher teach useful strategies explicitly?

Some scene management operations can be formulated as rules having a similar format to that described earlier. For example, consider the operation of adding a token in the following circumstance. One may have learned from painful experience that when shopping it is bad to ignore the price of the item one is buying. If in the Mind's Eye there is a token b for something you desire, you might want to allocate a previously unassigned token to represent its price. A simple extra rule would be of the form:

$$\text{ForAllTokens } b \, \{\text{Possible\_purchase} \\ (b) \Rightarrow \text{NewToken } a \text{ price\_of } (a, b)\}$$

Here "NewToken a" means that for a token a, for which the scene currently makes no assertions, add the assertion that follows "NewToken a." This operation would be done if there is a token that is a Possible_purchase, and the outcome would be to identify the price of that.

For the Mind's Eye to move on from processing one scene to processing another, we need a way of freeing up tokens by removing the attributes that apply to them. For this we can use, as a complement to the NewToken operator, an analogous operator called Remove. Invoking this would remove all attributes that apply to token a.

Belief Choice

The disciple Tzu-K'ung asked Confucius about government:

> The Master said, "Sufficient food, sufficient weapons, and the trust of the people." Tzu-K'ung said, "Suppose you had no choice but to dispense with one, which of these three would you forgo?" The Master said, "Dispense with the weapons."

Tzu-K'ung said, "Suppose you were forced to dispense with one that was left, which of the two would you forgo?" The Master said, "Dispense with food, for from old death has been the lot of all men, but a people without trust will not endure."[10]

The role of trust is indeed fundamental to all educable entities that receive information that they cannot verify for themselves. Accepting instruction from others puts us at risk of becoming a victim of unreliable and malicious agents. To flourish, an entity needs to have an effective policy regarding which sources to trust. By *belief choice*, I mean the policy used to decide when to accept a belief from others.

While discussions of human credulity often portray children as being highly credulous, experiments show that young children already use sophisticated methods for deciding whom to trust. In some such experiments, children interact with two individuals who give different names to previously unfamiliar objects. The child must decide whom to believe. In general, from an early age children would choose to believe a person they have a strong relationship with over a stranger.[11] At age three, they would still do this even if the familiar person has a history of having given names to objects that the child recognized to be wrong. By age five, though, they will give more credence to a relative stranger who has some history of having been accurate than to a familiar person with a record of inaccuracy. What happens if there are two strangers to choose between with similar records of accuracy? Then the stranger who comes from the same cultural group as the child, as determined by accent, for example, is the one believed. Experiments have also been done on adults, who, like children, use information about the provenance of theories to decide whether to adopt them. We tend to follow opinion leaders, people in authority, and those whose beliefs we share. Analogous experiments show that among chimpanzees, those with higher status are trusted more.[12] We all need ways of deciding whom to believe. Evolution has provided us and other species a start.

The biases humans have in choosing what to believe go beyond decisions about whom to trust. Psychologists have identified many other such biases. How we reason with the information we are given is not necessarily best described by formal logic, as shown by Johnson-Laird.[13] Much work has been also done, for example by Alison Gopnik, on how we attribute causality.[14] Also, there may be competition between our own experiences and what we are told. Humans integrate these two forms of belief acquisition so well that it can be difficult to tease apart what happens when they are in competition.[15]

Belief choice in humans is partially understood through the studies just cited. These biases are highly relevant in the fields of public persuasion and political propaganda. We will return to this in the section "Education and Propaganda" in chapter 12. Belief choice is a fundamental parameter of any educable entity.

CHAPTER 9

Educability as a Model of Computation

In this chapter I will first argue that the definition of educability is a plausible description of the essence of the human capability that I am seeking to capture. I will go on to argue that it is a robust model of computation.

Evolutionary Plausibility

An alien looking for so-called intelligent life on Earth ten million years ago might have found little sign of it and in its confusion promptly departed. One looking for "educable life," however, would have observed that most of the constituents of educability were already in place. Learning with generalization was being performed by long-existing neurons, and random networks of them were sufficient to implement symbolic naming. Many species had the capabilities of chaining. Scene analysis as captured by the Mind's Eye existed in mammals and more.[1]

The alien would have noted that the single step needed to reach educability was the appropriate integration with existing capabilities of the ability to acquire rules by instruction. It may have also noted that the already existing capabilities for imitation and teaching could serve as preparation for this elaboration. The suggestion made in chapter 3 that educability was a characteristic such that individuals having more of it had a selective advantage over those with less is therefore quite concrete in terms of the capabilities that animals had ten million years ago. Once the prerequisites were in place, as they were on Earth ten million years ago, educability's subsequent emergence may not have been a negligible probability event.

The alien returning in the present day might also find plausibility in the claim that the ability to acquire rules by instruction is fundamental to the newly evolved behavior. The evolution of this ability appears to have been accompanied by unparalleled desire for individuals to use it. This is manifest in humans both seeking and offering instruction even when totally unsolicited. Adults are willing to convey their thoughts to others somewhat indiscriminately, while children constantly ask questions of adults to an extent not seen in other species.[2]

The Model Is Explanatory

I said at the beginning that I was looking for a distinguisher that is also explanatory of our capability for civilization. We interact with the world and one another in disparate and discrete steps, many not in our control. We speak and hear others speak some words, we cast our eyes in different directions, and we have diverse physical experiences. We manage to organize the information we gather from these disparate activities so that we can respond with competence to the challenges the world offers. We would not be dominating the world as a species if we did not have such competence.

What is the explanation for how we achieve such competent behaviors? My answer is that the capabilities inherent in educability constitute the explanation sought. We interact with the world through some basic sensors and actuators, by means of which we can receive information from and make changes to our environment. First, we have a core capability of learning from experience and generalizing from it successfully. This constitutes a method of making sense of the many often disjointed experiences we have and finding regularities in them. Second, we can acquire explicit descriptions of good generalizations discovered by other individuals from their experience, so that we can enjoy the benefits without having to pay the costs of having those experiences. Third, we can apply these generalizations and combinations of them to individual situations. Further, all these capabilities encompass the ability to analyze scenes into their constituent parts and the ability to operate with arbitrary symbolic names. The probabilistic arguments in chapter 7 explained why in this system these disparate interactions add up or, in other words, are integrative. All this provides an explanation for how we can acquire effective responses to a complex world through a series of interactions that individually have much arbitrariness and are beyond our control.

Teaching to Reason

The previous section addressed the question of why educability captures the human capability that is responsible for our civilization. In the final two sections of this chapter, I will give two further, albeit less direct, arguments.

The Russian psychologist Alexander Luria went to remote villages in Central Asia in the 1930s and asked the Indigenous people the question: "All the bears in Novaya Zemlya are white.

Ivan went to Novaya Zemlya and saw a bear there. What color was the bear?" The response Luria typically got was "I've never been to Novaya Zemlya. You'll have to ask Ivan."[3]

Here Luria was interested in how humans reason when the premises presented are unfamiliar or false. Such work has been followed up in other parts of the world with similar results. Scribner and Cole repeated the experiments in West Africa and sought to distinguish the roles of literacy versus schooling in the ability to respond to such questions. They concluded that while two years of schooling helped considerably, literacy without schooling did not.[4]

Later work explored whether schooling as such was actually necessary. Dias and Harris asked similar questions to four- and six-year-old children in the United Kingdom.[5] They found that the answers the children without schooling gave were much improved if the questions, involving false or unfamiliar premises, were phrased so as to refer to a make-believe rather than a factual situation.

These results are consistent with the view taken here that humans have a basic chaining mechanism that is captured by the educability model, and that it can be harnessed to perform particular kinds of reasoning through a variety of interventions. In particular, while educability enables us to acquire a theory of a particular subject such as mechanics, it also enables us to acquire new methods of reasoning that are quite general. I shall mention some other methods of reasoning that, once learned, can yield better answers to everyday situations than do our uneducated intuitions.

A fundamental aspect of our civilization is the use of numbers. Many animal species have a sense of quantity whereby they can distinguish between small sets of objects according to their number. Humans, when taught, can manipulate exact numbers on a much larger scale. It is natural to ask how humans do this. Is having words for the numbers essential, or is the human capability an extension of the quantitative sense that we share with animals? We humans invest enormous energy into the teaching of basic arithmetic and associated reasoning about exact quantities. Much work has been done to elucidate how the human capability for reasoning about numbers can be enhanced.[6]

At a more advanced level, basic algebra is a powerful method of thinking about quantities. Assigning a symbol x to an unknown quantity and manipulating the known constraints on the quantity to obtain the value of that quantity is a useful reasoning method. It is unintuitive enough that even some who have been through formal education find this idea not so obvious. I knew one successful person who claimed that he had waited in vain throughout his high school days for someone to finally tell him what the value of x was.

An area where intuitions often fail is the estimation of probabilities. Suppose you are chosen at random to be tested for a rare disease and are told that the error rate (i.e., both when the test says *yes* for those who do not have the disease and when it says *no* for those who do have the disease) is 1 percent. Suppose the test result comes back positive. Should you be worried? The answer is, all things being equal, that if the disease is rare enough, say it occurs with probability $1/1,000,000$, then you should *not* be worried.

The following probabilistic calculation shows this. Of the four possible combinations of *having the disease* and *test outcome*, the only two you need to consider are those in which you test positive, as that is what has happened. Now the probability of not having the disease is $1-1/1,000,000$, and the probability of testing positive if you don't have it is $1/100$. Hence the probability of not having the disease and testing positive is the product of these two probabilities, $(1-1/1,000,000) \times 1/100$, which is about $1/100$.

On the other hand, the probability of having the disease is $1/1,000,000$, and the probability of testing positive if you have it is 0.99. Hence the probability of having the disease and testing positive is the product of these two probabilities, $(1/1,000,000) \times 0.99$, which is about $1/1,000,000$.

So, with regard to these two alternatives, you are much more likely to be in the population that does not have the disease but falsely tests positive than in the population that does have the disease and tests positive. (Naturally, these calculations need to be modified if more information is available, such as that you recently traveled to a country where the disease is rampant, or that you are experiencing symptoms of the disease.)

As this example shows, being taught the basic rules of probability enables us to reason in a powerful way about everyday situations that involve uncertainty. Among other applications: How does one judge whether a combination of events that has occurred was so unlikely that the possibility of mere coincidence should be dismissed?[7] How does one judge whether for two co-occurring phenomena, one causes the other?[8] These are challenging problem areas, and one can be taught methods of argument that help.

The earliest known systematic study of probability is believed to have been made as recently as the sixteenth century by the Italian mathematician Gerolamo Cardano. This is surprising since games of chance had been played for thousands of years. As far as we know, before Cardano, no systematic way was known for analyzing gambling or hosts of other everyday questions that we currently think of as probabilistic. For centuries, probability theory failed to emerge despite significant advances in other branches of mathematics. It is difficult to identify any precipitating event for which it needed to wait.

There are many other methods of reasoning that are currently commonplace but did not exist before someone invented them. Purposeful experimentation, as done in experimental science, relies on specific methods of reasoning for deriving conclusions. More generally, reasoning about hypotheticals, that is, considering thoughts without committing to their truth, offers useful avenues of argument. Considering counterfactuals, propositions that have not happened, and following through the consequences is one form. The paradigmatic mathematical proof associated with Pythagoras that the square root of the number 2 is not the ratio of two whole numbers starts by assuming the opposite and showing that that would lead to a contradiction.

We will see later in this chapter, in the section on "Universality," that one can mimic any computation in the Mind's Eye. This makes the notion of teaching to reason realistic. Each method of reasoning comes with specific rules. A generic approach to mastering a new reasoning method, such as the ones just described, is to mimic its rules in the Mind's Eye.

Less precisely defined than some of the above-mentioned reasoning methods have been approaches to improving an individual's judgment by having the individual reflect on their reasoning process. In arriving at an opinion, what were the assumptions made and arguments used? Socrates is associated with this movement in Western philosophy. "Critical thinking" is a more modern term and encompasses a variety of ideas that derive from it. Many believe that developing critical thinking skills should be an essential goal of formal education.

It is possible that the progress of our ancestors' cognitive activities tracked not only the invention of new artifacts but also the discovery and dissemination of more powerful methods of reasoning. This would require no biological evolution, and its development may leave little physical trace. It is conceivable, for example, that the evolution of reasoning methods across successive generations was initially slow. This may have been another limiting factor in the glacially slow emergence of civilization for an otherwise educable species.

Measures of Cognition

I will now observe that the numerical parameters of the educability model can be related to existing practices in the measurement of human cognitive capabilities. This provides an indirect argument of relevance for the model.

The word "intelligence" in all its many meanings encompasses a wide range of human capabilities. We flatter ourselves to believe that we outperform other animals in these abilities. As already discussed, the concept of intelligence has proven notoriously difficult to define, whether formally or informally. It has been proposed that "intelligence" subsumes multiple different competencies rather than one central one. Howard Gardner famously talked about "multiple intelligences." He identified the following eight: linguistic, logical-mathematical, musical, spatial, bodily/kinesthetic, interpersonal, intrapersonal, and naturalistic.[9] Several other notions of intelligence, such as emotional intelligence and social intelligence, have been suggested also as being important for individual human success.

Currently a widely used method for evaluating children's cognitive abilities in schools is the Wechsler Intelligence Scale for Children, or WISC. This consists of several distinct subtests that combine to form distinct indexes. The five basic indexes are Verbal Comprehension, Visual Spatial, Fluid Reasoning, Working Memory, and Processing Speed. In other words, the current practice of evaluating children goes well beyond measuring a single capability. It involves measuring several parameters that are each associated with a different identifiable capability. These five capabilities appear to have little in common. Indeed, the indexes were designed so as to measure substantially orthogonal abilities. They can be combined to provide more general measures, which can be further combined to produce one overall so-called IQ score.

One can try to relate these separate indexes to the educability model. The Working Memory Index measures short-term memory capacity and corresponds to the number of tokens in the Mind's Eye. The Processing Speed Index measures the time taken for computing basic tasks, which is a parameter of any computing device, including the ELS. The other three all involve reasoning, which in the ELS corresponds to the application and chaining of rules.

The effectiveness of the rule-manipulating infrastructure in our brains may not be measurable directly. Indeed, it is probable that there are separate components involved in the various kinds of reasoning. The section on "Innate and Acquired Attributes" in chapter 6 discussed attributes implemented at birth. The reasoning capabilities that use these attributes may depend on these different implementations, which would give rise to such separate indexes. The relative capabilities for processing verbal versus visual inputs may vary between individuals. The WISC test, through the Verbal Comprehension and Visual Spatial indexes, may be indirectly measuring parameters relevant to these different capabilities.

The ELS formalism is an attempt to describe the functionality of educability that I believe humans have. An analogy would be Turing Machines, as they describe the functionality of digital computers. The realization in both cases is indirect and involves some parameters. The correspondences between the numeric quantities in the model and the realization are sometimes clear. For example, the amount of memory used in a Turing Machine and a computer when executing corresponding tasks is approximately the same. Similarly, some of the indexes measured by the Wechsler Test are recognizable as relating to the quantitative parameters of the ELS.

This does not mean, however, that this or other existing tests designed to measure "intelligence" provide a good measure of educability. Existing tests might measure certain parameters of educability but not the main capability.

Universality

Universality is an indispensable feature of digital computation. Computers are designed not for just one specialized task, but for all computable tasks. Turing's universal Turing Machine is the origin and ultimate demonstration that such an ambition is realizable. In the early days of computers in the 1950s, it was not obvious to all computer designers that universality was the way to go. Some argued that scientific computations were different enough from commercial data processing. Companies therefore made different machines for these different markets. The advantage of universality is that you only need to make, buy, and maintain one machine. It was not entirely obvious that this advantage outweighed the possible efficiency advantages offered by more specialized machines.[10] In the decades since, universal machines have generally triumphed, although there has been ample room for auxiliary special purpose circuits also.

We humans do not have the option of buying and maintaining more than one brain. It would be helpful if our brains were computationally universal. Fortunately, in a certain sense they are. We are all able to memorize a more or less arbitrary sequence of mental instructions and apply them in our head, up to the quantitative limits of our memories. Such a capability is a prerequisite for educability, which requires the ability to apply the rules we have acquired by instruction.

The primary challenges of executing a sequence of instructions in one's head, such as when evaluating the value of an arithmetic formula, are (i) memorizing the sequence; (ii) keeping track of which one of the instructions we are currently executing; and (iii) keeping track of the current values of the variables we have computed. For example, if the last instruction is to add 2 and 3, then for (iii) the value 5 needs to be stored somewhere. An ELS as defined previously is up to these challenges. The basic strategy is (i) to represent each instruction as a rule; (ii) to assign a separate token to each instruction and have a special attribute for each to indicate whether that instruction is the one being currently executed; and (iii) to assign a separate token to each variable and have a separate attribute for each possible value to indicate what the current value is. The notes for this chapter give further details of how an ELS can simulate this universality.[11]

Any ELS realization of this capability will have some numerical parameters, such as bounds on the number of tokens and instructions. Just as with machines, it becomes truly universal only if these parameters can be made arbitrarily large. For example, a human executing a complex calculation may need access to an unlimited number of sheets of paper to record the intermediate values of the variables as the computation proceeds if these are too many to hold in memory.

This ability to represent and evaluate *any* set of rules in a class is different in kind from the ability to evaluate a specific learned circuit with physical sensor inputs and actuator outputs, such as that of the sea snail. In the case of the snail, we believe, the physical world it experiences through its sensors determines the processing in the fixed way defined by its circuit. In contrast, the ELS can execute any set of rules it receives by instruction up to its resource limitations.

The Educability Model Has Robustness

In chapter 5 model robustness was defined as the requirement that the meaning of the model should be the same for variants of the model that have the same intent. The archetypal instance of this is Turing computation. All attempts at defining models with the same intent, namely, capturing general computation, have yielded none with more power than Turing machines.

As pointed out earlier, there is also a second sense in which one can claim robustness. The intent of a computation may be a specification of the outcome that is to be achieved. For example, one's intent may be to have a program that puts names in alphabetical order. There are many different algorithms for realizing this task. These algorithms would perform different sequences of steps for the same input and may vary in the quantity of resources of time or memory that they require. What is sought is that every computer should produce the same result when given a set of names to put in alphabetical order, even if their algorithms differ. In other words, a characterization of *what* an algorithm achieves, beyond the step-by-step description of *how* it achieves it, provides an important second kind of robustness.

The opposite of robustness here is arbitrariness. A procedure for which one does not specify either a description of what it achieves or alternative ways of achieving whatever it does achieve would be arbitrary. In this sense, an individual computer program may be arbitrary and lack robustness.

What we are claiming for our ELS is that it has model robustness in *both* senses. The first sense is invariance under change in model, which we have called type A. The second sense, which we called type B, is having a specification of the outcome, other than the step-by-step description of how one gets to it.

PAC learning has Type B robustness since there is a definition of what PAC learning achieves, namely, accurate predictions with efficient error control. For the step-by-step execution of instructions explicitly acquired, and for chaining, the desired outcomes are also clearly defined. Hence all three aspects of educability have a clear Type B style definition of the desired outcome. Models defined in terms of vaguer notions would not have this robustness.

PAC learning also has some Type A robustness. As remarked earlier, PAC learning remains unchanged under a variety of changes in its definition and is therefore robust in this sense. Also, if C^*, the rules acquired by instruction, have the same power as Turing machines, then by the Church-Turing Thesis described in chapter 5, this component of educability will inherit Type A robustness also.

Evidently, as models of computation, the natures of Turing Machines and Educable Learning Systems are different. The former has a much simpler definition. One commonality they share is that physical realizations of them are constrained by quantitative resource limits. For any physical realization of a Turing Machine, the main limitation at any instant is that of memory size, that is, the space used for storing the intermediate data. In principle, it can use as much memory as its computation requires, though at any instant it has used just some finite amount. Another parameter involved is the processing time of a single step. Note that while this speed parameter influences only how fast the Turing Machine obtains its solutions, any space limitation does influence what can be computed, since there are computations that are arbitrarily hungry for space.

An Educable Learning System is similarly limited, but now there are many more parameters. As one defines learning phenomena of more ambitious kinds, the model obtained necessarily gets less elegant. An ELS is indeed a more complicated notion than a Turing Machine. Any physical realization will have several *numerical parameters*, such as the maximum numbers of rules and tokens allowed and the time for executing a basic step. There are also some *nonnumerical* parameters: (i) the class C of rules that can be learned and the class C^* that can be acquired by instruction, (ii) the scene management rules, and (iii) the belief choice policies.

Despite all these extra parameters, the ELS has computational robustness in the senses previously described. This robustness suggests, for example, that future models that aim to capture similar phenomena will be relatable to it.

CHAPTER 10

Promoting Educability

Educability as an Individual Capacity

In young children generalization is a spectacular phenomenon for all to see. After seeing depictions of giraffes in books, a child will easily recognize them at a zoo or as a statue in a garden. Generalization is a cognitive skill that is surely critically useful throughout life. Curiously, this skill is rarely noted for individuals. One rarely hears the accolade, "Oh, he is so good at generalization" or "Why, she is so inductive."

Equally, one seldom finds remarks on this ability in literature or history. An exception in antiquity occurs in the commentary on the life of Alexander the Great written in the second century by Arrian of Nicomedia. Alexander was successful in competitive exploits. He had received exceptional education from private tutors. What were his qualities that might have contributed to his success? Arrian, in a short chapter on "The Character of Alexander" in an otherwise lengthy text, remarks that Alexander "was very clever in recognizing what was necessary to be done, even when it was still a matter unnoticed by others; and very successful in conjecturing from the observation of facts what was likely to occur."[1] This capacity may have been evidenced already in his childhood when, witnessing an apparently un-rideable horse, Bucephalus, being offered for sale to his father, Alexander correctly conjectured from observation that the horse's root problem was fear of its own shadow.[2] Alexander's father reluctantly purchased the horse for him, and the rest, as they say, is history. Because of his age, Alexander most likely came to the situation with less knowledge of horses than some others present. Nevertheless, he was able to make better use of the observables of the moment.

Besides generalization, educability also involves the ability to learn explicit theories from others and to apply them. There is the notion of someone being a "quick study," which usually refers to the ability to quickly absorb a set piece of material, as may be required in a new job. There is also the concept of a person being "coachable," implying that they can be trained to perform a particular task. Educability is more general than these latter notions in that it implies that the knowledge gained will be useful for purposes not foreseen at the time of the study or the coaching.

There are many other terms we use to recognize valuable abilities that somehow involve learning and are not well captured by conventional notions of IQ. We all know the term "street smarts," often used to refer to an individual who has an uncanny ability to negotiate the practicalities of life. Or consider the accolade that someone has "common sense." The concept of being "well-educated" is relevant. It usually refers to someone having learned a large amount of knowledge across a broad range of areas. If one has the benefit of formal education for fifteen or more years, it may be easy to emerge well-educated. Educability is about the capacity to take good advantage of whatever educational opportunities arise, whether formal or informal. Persuasive evidence of the power of educability is offered by the example of individuals who lacked those many years of formal education but who, through self-education, became well-educated. William Shakespeare held his own against the other prominent playwrights of his day in England, most of whom had a university education, which he apparently did not have.

One would expect that high educability is useful for every occupation. One can only speculate as to which occupations demand the most. For instance, our political leaders often need to make judgments on matters well beyond the knowledge and experience they had when elected, such as those arising from new technologies. Sometimes we praise candidates for high political office as having the quality of curiosity. Perhaps we should raise our sights and demand that our leaders be able to act out their curiosity and be highly educable, irrespective of whether they are highly educated.

On the other hand, as has already been said, educability does not encompass all valuable human capabilities. Any predispositions to be empathetic, ask good questions, tell appropriate jokes, be creative, and recognize the face of an individual one has not met for many years may also be useful. We are not aiming to describe the totality of the human condition. Educability seeks to articulate a core capability, one that accounts for our success as a species and that all humans share.

Educability of a Population

If we succeed in identifying educability as the Civilization Enabler, can useful applications follow? Well, if it is our defining characteristic and is responsible for our civilization, then we may want to promote it. An important question would be: Can the educability of an individual be enhanced by education or training? An affirmative answer here could have major consequences. It would allow us humans to play even more to our strengths. Such an affirmative answer seems likely simply because almost every other human capacity can be enhanced.

Which methods can enhance educability? It may be that some current methods already succeed. If we pursued educability as a goal more purposefully, could we have more success? The limitation to date has not been that the goal of improving the effectiveness of education has not been pursued. It is more perhaps that the goal has not been articulated in quite such concrete terms.

Another area to explore would be whether a critical period exists for educability after which our capacity for educability is diminished.[3] We know that lifelong learning is possible, but it may be harder to learn some things as we age.

There is much current discussion of the increasing pace of technological change and its consequences for employment. Individuals who may have had a single career in the past may now need several career changes in a lifetime. Governments rightly recognize the need for more educational resources for retraining. Inseparable from this discussion, surely, is the educability of the population, namely, its readiness to take advantage of the educational resources made available. I suggest that it is the responsibility of governments to promote the educability of their citizens as much as it is to safeguard security, promote good health, and provide education.

One can make an analogy between health and educability. It is one of the responsibilities of governments to promote public health. This involves having measures of health. These measures relate to individuals in that they measure how many individuals have various diseases. But they do not necessitate governments ranking their citizens by an individual health score and distributing rewards accordingly.

I would argue *against* basing decisions that are life-altering for individuals on a test of educability or any similar single test. Skepticism about the value of individual testing revolves around the problem of making tests that are fair to individuals with divergent backgrounds. As will be discussed in the next section, the aim of an educability test would be to test only for material learned after the beginning of the test and nothing else. This comes close to the goal of creating a fair test but is not synonymous with it. The problem is that the competence exercised during the test, namely, educability, might have been enhanced by earlier life opportunities. Hence, even if perfect tests for educability could be developed, this will not solve the fairness problem.

A further issue is that a test is usually imperfect as a measure of the targeted human skill. Scores on almost any test can be improved by systematic test preparation that does not necessarily enhance the candidate's abilities in the tested skill but merely enhances the test taker's ability to take the test. Unequal access to test preparation services can make tests unfair. It would appear likely that educability can be usefully compared among individuals using tests for which candidates are not previously trained, but not by tests for which some have been trained and some not. In this case, it would be better not to distribute rewards, such as college admission, based on performance on educability tests so that there is no incentive to train for them. Then these tests would remain valid for the societal purpose of enhancing educability for all.

The societal purpose of measuring educability would be to evaluate the effectiveness of various interventions on educability. For a proposed intervention, one would be looking for how much the measured level of educability has changed after an intervention from before.

Measuring Educability

Let us consider the possibility of designing and validating methods of measuring educability. Most simply, one would need to measure competence in the three pillars of educability, namely, (a) learning from experience, (b) being teachable by instruction, and (c) applying knowledge one has acquired by either means. The distinguishing characteristic of an educability test would be straightforward: it should test only for knowledge *acquired during the test or deduced during the test from that knowledge.* Knowledge possessed beforehand should give no advantage.

The challenges in developing such measures should not be underestimated.[4] As a warning, consider Luria's question to the villagers about the color of the bear, as related in chapter 9. One might have thought that Luria had posed the perfect educability test to the villager, in that the information needed for the correct answer, the color of the bear, was provided explicitly in the question.

In chapter 9 I noted some correspondences between existing tests designed to test intelligence and some basic features of educability. These current tests measure, among other things, basic quantitative parameters such as the speed of response and the capacity of short-term memory, which indeed are key factors in the performance of an ELS. Others measure reasoning abilities, which are also highly relevant. These current tests, however, do not set out to evaluate an individual's educability.

Nonetheless, some questions on existing tests already test for some of our three pillars. For instance, if one is given three examples of a concept and is asked to choose which one from a set of others is also an example of the same concept, then this would test (a) and (c). The concept may be red rectangles, and the examples may be red rectangles of varied sizes in different orientations. The examples to select from may be a variety of shapes of different colors. The examples that meet the test criterion will be those that are red rectangles.

As far as testing (b) in combination with (c), the questions would need to be more elaborate. They would describe a theory, present a situation, and then ask what conclusion could be reached by applying the theory to that situation. The theory should not be a theory already known to the student, since then one is testing for prior knowledge. Existing standard tests that come closest are tests of comprehension. In these, some paragraphs are presented, and questions are asked about them. For educability, the requirement would be that the paragraph present something new to the student, so that only knowledge gained during the test is evaluated.

Existing test questions often test for previous knowledge. For example, questions such as, "If apple is to eat then coat is to . . . ?" test one's understanding of the relationship between "apple" and "eat" and present no new knowledge to be absorbed. Drawing a conclusion from one example tests for an individual's prior understanding of a concept and one's "similarity metric," as discussed at the end of chapter 4. There is nothing wrong with testing for prior knowledge or the similarity metric, but it is different from educability.

Comprehension tests that evaluate whether one has understood the vocabulary or has knowledge of the concepts to which the text refers are also testing for prior knowledge. To make them tests for educability, the text needs to include content that is unfamiliar to the student. It is the acquisition and use of this new content during the test that is to be tested. This content can be real or fictitious. If it is not invented for the test, it should be obscure enough that it is unlikely to be known previously by the student.

There are challenges to realizing educability measures. Every effort will need to be made to minimize the effects of prior knowledge on students' understanding of questions. Nonverbal formulations may help. Digital technology offers a wide range of possible formats. For example, comprehension-style tests appropriate for measuring educability may use the medium of videos. Also, digital technology offers new opportunities for learning by interaction. As discussed at the end of chapter 4, interactive learning, in the forms of play and exploration, occurs in many animal species including humans. Hence educability tests could include tasks where the student acquires information through interaction, perhaps in a computer game, or by querying a search engine on the web. In all these cases, appropriate familiarity with the digital format would need to be ensured since the capability that is to be measured would still be that of acquiring new theories and applying them.

One can envisage academic courses given over several months also serving to measure educability. The courses would realize the three pillars of educability in an integrated way, so that knowledge learned from examples and acquired by instruction would need to be applied in combination. Courses that emphasized just one of the three aspects—learning from experience, acquiring rules by instruction, or case studies—would not be appropriate by themselves. A clear advantage of measuring educability via a course is that the course is likely to be an activity of value in itself to the student, while test taking rarely is.

By making the sharp distinction that one wants to test for the absorption of new knowledge provided during the test as opposed to testing for knowledge already in the student's head at the start, one is defining a new measurement goal.

Long-Term Retention

I have been emphasizing that essential to education, as opposed to training, is the capability to apply acquired knowledge in circumstances not foreseen or foreseeable at the time of the acquisition. The application may involve several pieces of knowledge, acquired at widely separated times and in entirely different circumstances. It is implicit in this statement that the pieces of knowledge may need to be retained for some time. Without that, the chances of combining many pieces acquired at different times will be severely restricted.

In the Educable Learning System, as described, knowledge is retained without time limit. While rules can be updated as deliberate acts of learning or instruction, there is no mechanism for automatically forgetting rules. The model retains rules indefinitely, or until a deliberate act of updating.

In humans the testing of knowledge retention is challenging. Formal education usually does not evaluate retention well—it is often enough to acquire knowledge just before it is tested. It is not clear for how long humans retain information that they have acquired immediately before they are tested.

Psychologists have given some attention to the question of how best to ensure that taught material is retained for a long time. It is believed, for example, that *retrieval practice*, or repeated recall, is an effective technique.[5] However, measuring long-term retention is difficult in general. It is a dimension of educability measurement that needs to be addressed.

CHAPTER 11

Artificial Educability

Fear of Technology

In 1863 Samuel Butler predicted that machines would take over and "man will have become to the machine what the horse and the dog are to man."[1] His argument was not based on any specific technology. It was that humans find it impossible to resist adopting the most advanced technologies available and in that sense surrender control of their lives to technological change without taking into account their own human interests.

Another concern having a long history is automatic decision-making. If a country sets national examinations and its universities are bound to admit students strictly according to the exam results, then the admission decision can be considered to have been automated. Will the outcome be fair given that individuals start with varied advantages? Will the overall outcome, namely, the type of students who are admitted, be aligned with the country's interests? These questions of *fairness* and *alignment* are especially alive today now that computer technology is widely used for decision-making about individuals. The discussions they give rise to are often not specific to any one technology.

Earlier chapters have described a specific approach to human cognition. This chapter will focus on how this approach provides an understanding of current and future digital technology. It is suggested that a scientific understanding will be needed for us to cope successfully with the opportunities offered and risks posed by technologies that can learn.

The Imitation Game

The first two papers that described artificial intelligence in recognizably modern terms were both written by Alan Turing.[2] The first, written in 1947–48 and entitled "Intelligent Machinery," was dismissed as a "schoolboy's essay" by his boss. It remained unpublished and little known until the 1980s. It was, however, the basis of his second paper, "Computing Machinery and Intelligence," which he published in 1950 in the journal *Mind*. This work contained the Turing Test and defined some of the issues that continue to be debated to the present day.

Turing was concerned with what it meant for a machine to be "thinking" and how to make machines that could think. To the former question, his answer was simply that if the behavior of the machine was indistinguishable from that of a thinking person, then we should consider the machine to be thinking. If there are no detectable behavioral differences, there is no more basis for rejecting the machine as a thinker than there is for rejecting your neighbor as a thinker.

The second question, how to realize such a thinking machine, has turned out to be highly consequential. Turing raised and discussed the basic conundrum: Should one program the machines, or should they learn? He advocated a combination where, first, one programs a machine to realize the capabilities of a child, and then in a second stage one gets the machine "educated." In his words: "We have thus divided the problem into two parts. The child-programme and the education process." In this paper Turing has a section called "Learning Machines," and in the earlier paper, a section entitled "Education of Machinery." He does not make a technical distinction between learning and education. Nevertheless, in looking back, his use of the word "educate" is striking, if only because it has been so rarely used in the artificial intelligence literature since.

Turing may have intended "Computing Machinery and Intelligence" as a provocative position paper regarding a topic close to his heart, rather than as a scientific contribution. Indeed, the paper has served the former role over the seventy years since. Unlike Turing's paper on computation, published in 1936, this paper did not trigger a sequence of specific mathematical, scientific, or engineering developments. He used the words *thinking* and *intelligence* but did not try to relate or define them. He says explicitly in his first paragraph that such terms are difficult to define: "The definitions might be framed so as to reflect so far as possible the normal use of the words, but this attitude is dangerous. If the meaning of the words 'machine' and 'think' are to be found by examining how they are commonly used it is difficult to escape the conclusion that the meaning and the answer to the question, 'Can machines think?' is to be sought in a statistical survey such as a Gallup poll. But this is absurd."

He then puts forward the Imitation Game, which treats thinking as a "you know it when you see it" phenomenon. He asserts that one should consider a machine as capable of thinking if a human remotely connected to both it and another human cannot reliably distinguish between the two by asking them questions.

The Imitation Game *is* compelling if one identifies "thinking" with the totality of the human mind. Indeed, a later radio broadcast by Turing in 1951 hints that the human mind was the focus of his interests: "It might for instance be said that no machine could write good English, or that it could not be influenced by sex-appeal or smoke a pipe. I cannot offer any such comfort, . . ."[3] The goal of his machine would be to behave like a human, rather than to realize a capability that one would need to define in other terms.

Beyond the Imitation Game

The view taken here is that for making further progress both in understanding humans and in improving technology, one needs to *define* the functionality one wants to realize by machine and not treat it as a "you know it when you see it" phenomenon. This is what Turing resisted in the context of "thinking." He rightly said that defining such a term by common usage does not work. The approach taken here, instead, is to formulate the phenomenon of educability as a Robust Model of Computation, exactly as Turing did for computation in his paper in 1936 but did not do for thinking or intelligence.

The decades since the publication of "Computing Machinery and Intelligence" have seen much progress in computer science. Informed by developments made in this period, we are now in a much better position to *define* computational phenomena and functionality. In earlier chapters I provided such a definition for educability. If educability *is* the capability that made human civilization possible, then it would seem reasonable to seek to exploit this capability also in technology. There should be no overwhelming impediment since educability is being defined in terms of a computational model.

Almost everything we have asked computers to do can be done and had been done previously by humans. The scale and speed of machines may dwarf those of humans; however, the tasks that computer technology performs for us broadly reflect human interests and ambitions. If educability is basic to human cognition, then any new insights it provides should be useful for extending the capabilities of technology. As such, educability is not purely a study of humanity. One can expect that a better understanding of it will also translate into better technology.

It is widely believed that as far as computation itself, current computers have hit the maximum capability possible with Turing computation, putting aside efficiency considerations. There is no alternative known with greater computational capabilities.

It is equally possible, but certainly not proven, that with the notion of educability described here, human evolution has hit the maximum for this kind of capability for individuals subject to biological constraints. The most basic such constraint is limited lifespan. Educability is a mechanism that compensates by having many individuals in a community, efficiently learning, communicating, and sharing knowledge.

To clarify, I have been suggesting educability as just *one* aspect of humanity, though the defining one as far as our capability for creating civilization. Humans have numerous other aspects, many shared with other animals. I am not trying to define humanity or the human condition here, or to simulate a human. I am trying to isolate one critical capability.

This is different from Turing's "Computing Machinery and Intelligence" viewpoint. Turing's impact on how popular culture views intelligent machines is currently pervasive. For example, the film *Ex Machina* (2014), which references the Turing Test, is about a machine that was kept imprisoned after it was constructed. In the film, the machine's creator declared that his machine had passed the "AI Test" because it had made a plan to escape and in doing so had shown "imagination, sexuality, self-awareness, empathy, manipulation." Such lists of diverse human characteristics are more like describing the human condition. The question of whether machines can have them is appropriate for speculation, but such discussions have yielded few insights into making machines with more powerful capabilities.

AI

The term "Artificial Intelligence" is associated with a conference held in Dartmouth College over a summer month in 1956 and particularly with John McCarthy. However, neither that conference nor the research community that evolved from it ventured to define the term "intelligence," whether artificial or otherwise. Neither, as we have seen, have the psychologists, nor anyone else, with much success.

It is therefore not surprising that the term Artificial Intelligence has undergone some changes in usage. For the first several decades, it meant machines having cognitive competence with a high bar. The joke was that tasks that scientists managed to get computers to perform ceased thereafter to make the cut, so that true AI was forever over the horizon. The term "AI" was reserved to describe whatever was impossible for machines to achieve at the time. Successful advances in the use of computers were categorized as computer science rather than AI. It was noted that the field of AI was unusual in being named for its aspirations rather than achievements.

With the entry of industrial players into machine learning, the terminology flipped. Suddenly, sometime in the early 2010s, tasks that computers *could* solve came to be called AI, and the problems computers could not solve became less talked about. Also, it became commonplace to refer to a computer running an AI program as "an AI." Thus, a little unusually, an object was named for the aspirations of its designers rather than for what it was. In this way, AI became largely synonymous with what machine learning algorithms were achieving.

There were good reasons for industry's sudden interest in machine learning. Previously, one heard from industry that machine learning products "do not ship" because they do not have predictable enough behavior. Several things changed. First, around 2006 the cost of computing suddenly became many orders of magnitude cheaper with the advent of general-purpose graphics processing units (GPUs). Enormous computing power, with support for programming, suddenly became available at a moderate price. Second, by that time it was also possible, with the help of the internet, to collect and distribute large datasets. The internet also offered a feasible way of getting millions of examples, of images say, correctly labeled by humans. New services had become available that enabled humans working remotely around the world to be paid for this work. A third concurrent trend was that internet companies were in possession of copious amounts of data about which "probably approximately correct" predictions were good enough and useful. Detecting spam in email, ranking internet searches so that the most popular ones came at the top of the page, language translation, and analyzing the behavior of online shoppers were early applications. The capability for making enormous numbers of predictions, each of possibly modest accuracy, was suddenly useful. Small improvements in prediction accuracy improved user experience and increased profits. Critical to the wide adoption of machine learning for these applications was the fact that not much was at risk in any one prediction—no one was killed by a bad prediction.

With large datasets and cheap and abundant computer power, the question arose as to how best to exploit the sudden affordability of large learning effort, measured as N in chapter 5. Problems of particular interest were those of categorizing images, interpreting spoken sentences, and processing text. The answer found was that the backpropagation learning algorithm (rebranded as "deep learning") outshone its rivals on many such datasets.[4] A dataset that was particularly influential was ImageNet, which consisted of more than a million images classified into categories by humans working remotely.[5] The results of an annual competition that compared the performance of machine learning algorithms on this dataset proved decisive in the wider adoption of deep learning.

The following is a simplified description of the deep learning algorithm in the context of image classification. As inputs, the algorithm takes, say, $1,000 \times 1,000$-pixel images, where each pixel is represented as two numbers that describe its brightness and color. Consider a circuit with these two million inputs that performs some arithmetic calculations internally on these inputs and then outputs a single numerical value x. The intention is to train this circuit so that there will be a high probability that whenever the picture contains an elephant, the value of x is more than 0, and when it contains no elephant, the value of x is less than 0. The training consists of feeding the circuit with many images, some representing elephants and some not, and modifying the circuit so as to improve its power to classify. For instance, when an example is presented that the current circuit classifies incorrectly, the circuit is very slightly modified so that the output changes in the right direction on that input. If on a positive example of an elephant the circuit incorrectly computes $x = -3$, then the circuit is changed so as to increase the value of x. If on a negative example the circuit incorrectly computes $x = 2$, then the change would be such as to decrease x. In general, one might expect that while each update makes sense for the particular example that causes it, together the many updates may not, since any change in the circuit driven by one example may be in the wrong direction for other examples. It turns out that on the natural datasets arising from vision or speech recognition, these updates can be made to work so that in their totality they tend to improve the predictive ability of the circuit.

In earlier decades, it had been observed that while the rate of improvement of this algorithm can become imperceptibly slow as it is trained more and more on a given dataset, the algorithm rarely gets stuck. The more recent finding is that if one continues this training for long enough, long after the training set has been correctly classified, then impressive accuracies can be obtained in predicting on examples not seen during training. This algorithm had been found to be inferior to other known methods on the smaller datasets and the smaller computer budgets that had been available in earlier decades. With some of the large datasets currently available, however, and at the current much reduced price of computing, it has been found rewarding to expend enormous resources, perhaps millions of dollars, to train on a large dataset.

Some of the more striking applications of machine learning are those in the classical sciences. One area is that of identifying the three-dimensional structure of proteins from their chemical specifications. The question is to determine how these string-like chemical structures fold up in three-dimensional space given the physical interactions among their constituent atoms. Their folded shape determines the interactions they will have with other molecules and hence their function in living organisms. This folding can be determined, in principle, by simulating the physics forward, but this takes too many operations to be currently feasible for the accuracies needed. However, a machine learning system called AlphaFold, trained on molecules with known structure, has been shown to outperform earlier methods based on simulating the physics.

Such applications to the classical sciences are pleasing because they are gifts provided by machine learning beyond the original motivations of the field. The original motivations were the cognitive systems of living organisms, such as vision, speech, and language. The Church-Turing Thesis, discussed in chapter 5, implies that digital computers can compute, and therefore also learn, whatever the corresponding biological systems can compute and learn. The application mentioned to protein folding is still to do with computations performed by the physical world but is a little different because it generates the output in an entirely different way from nature.

The current successes of machine learning on cognitive tasks are in some sense unsurprising since nature already realizes them by some kind of learning process. But there is news here, too. Machines can now outperform humans in efficient generalization in many areas in which humans are good, including the classification of images. For example, machines can reliably classify images of human retinas according to gender, a task that human practitioners had not even realized was possible.[6] Machines can also classify microscope images for some medical applications far more reliably than can humans.[7]

Despite these successes, it seems clear that current efforts in AI fall short in realizing some major facets of cognition. They fall short in areas where one needs reasoning and the use of background knowledge. This is where I believe the educability notion has something to offer. This chapter therefore proposes to amplify the reach of machines in performing cognitive tasks by adding aspects of educability into such systems. The current trend in using supervised learning alone with larger and larger N will continue to yield better and better performance, and I expect that nothing will stop this pursuit, but I suggest that enhancements of the architecture toward educability will yield benefits that simply increasing the scale of current practice cannot attain.

The following is a metaphoric way of phrasing the conundrum. You may know something about appendicitis because you know three people who have had the disease. But if you need to know enough about the disease to make an important personal decision, you will go to an experienced doctor who has seen a thousand cases. A computer that uses an algorithm that behaves as a PAC learning algorithm will make better and better predictions the more examples it has seen. If it has seen a million cases, a situation well beyond individual human experience, then it may make predictions that are stunningly better than our intuitions can even comprehend. Some of the more spectacular applications of machine learning that use even larger datasets are the so-called *large language models* like ChatGPT, discussed in the next section. They perform beyond our intuitions in generating beguilingly smooth prose not because what they do is beyond our experience, but because they do it on a scale beyond our experience. They are trained on *billions* of sentences.

We are used to computers processing mathematical or logical tasks at or beyond human performance. Each time a new task that was previously the preserve of humans, such as language translation or the generation of smooth prose, is shown to be doable by machine, some express surprise and concern. But there should be nothing surprising in this. A distinction can be made between tasks that are *theoryful*, such as logic and arithmetic, where we understand exactly *what is to be done*, and the *theoryless*, where we humans can recognize when the task has been performed well, such as the generation of smooth prose, but *cannot formulate the exact requirement that has been satisfied.*[8] The argument has been made that supervised learning provides a process by which *certain* theoryless tasks can be performed, by humans or machines, as effectively as theoryful tasks given an appropriate training set. We need to accept the enormous successes of machine learning in such theoryless but empirically learnable tasks as being within the realm of the understood and not the mystical, and be less surprised or upset by it.

As a personal note, this author is, of course, delighted that machine learning systems with enormous powers of generalization have been realized. This confirms that PAC learning describes a powerful phenomenon that is technologically realizable. When the author formulated this notion in the early 1980s, no technological demonstration of this was available.

At that time, however, it was already clear that this would not be the last word on understanding cognition. The current book is an attempt to fill that gap. The reason we go to doctors is not just that they have seen a thousand cases of a disease. It is that we believe doctors deliver further results. Educability aims to capture this extra value that human doctors offer beyond merely having seen many cases.

Large Language Models

Before going on to a broader discussion of current AI, I will discuss Large Language Models (LLMs), such as GPT, a product of current AI that has attracted much public attention. What are LLMs? In the simplest terms, they are supervised learning systems that have been trained to predict the next word from a piece of text. When they are asked to predict words successively, they can produce arbitrary amounts of text. To be more precise, what LLMs predict one at a time are usually not words but word fragments. (These fragments are called "tokens," which are not to be confused with our use of that term here in the Mind's Eye.)

The predictive accuracy of LLMs comes from the amount of data on which they have been trained, often in excess of a billion sentences, and the amount of computation time expended, often costing millions of dollars. Ample evidence shows that their behavior is consistent with the requirement of PAC learning that the errors in their predictions decrease as a fixed power of the effort invested. Training on more data is good. Computing more on the same data is also good. Further, the gains in accuracy obtained can be well predicted by a power curve such as figure 2 in chapter 5.[9]

LLMs may be the largest existing machine learning systems, but beyond that there is no special mystery to them. The training algorithms currently used are transformers, a variety of the deep learning algorithm already described.[10] They are effective for the kind and size of data involved, but they have no intrinsic properties that make them exceptional among the practical machine learning algorithms that are known.[11] The accuracy of these LLMs depends on the vastness of the dataset on which they have been trained. To make the system more useful, subsequent levels of training and tuning that involve human testers are performed to achieve further goals. These goals may include responding to common question types appropriately, avoiding giving dangerous advice, and adopting a polite tone. The human testers may be asked to provide examples of questions and appropriate responses to them, or to choose between alternative responses given by the system.

LLMs are trained to predict the next word or word fragment, but that is not what makes them so intriguing. They generate well-constructed sentences that keep on topic for long enough that they remind us of human trains of thought. They are an example of *generative* AI in that many bits are predicted, in this case text, and the totality of the many bits has internal consistency.

Do these systems reason? We can try to answer this in terms of our earlier discussions.

The first thing to note is that the question of whether reasoning is useful at all when learning is done on a large enough scale is already nontrivial. It was the subject of the earlier section "World without Reason." There it was argued that *sometimes* the need for reasoning exists only because the learning needed to replace it would have to be on an impractically large scale. Currently, when the performance of these LLMs can be improved simply by devoting more computational resources, even to an unchanging training set, there is some incentive to just keep increasing the training data and training time and see what happens, rather than change the system to incorporate reasoning explicitly.

But can LLMs reason? Our earlier discussions of reasoning were in terms of chaining. The Robust Logic model describes a reasoning capability that I believe humans have. The primary mechanism of LLMs, PAC learning for predicting the next word, is clearly less than that. It does have one feature, however, that provides a *restricted* form of chaining. The feature is that as it produces its stream of words, it adds this stream to the input from which it generates the next word. This does make possible a chain where each word influences later words, and a long chain can be threaded through the multiple words it generates.[12]

The Robust Logic analogy would be that the LLM is using the words it generates, laid out in a sequence as text, in a similar way to how Robust Logic uses tokens in its Mind's Eye. Its utterances depend on its analysis of a situation as far as this analysis is possible from the words previously uttered. In this sense, LLMs have chaining capabilities, but of a particular restricted nature as compared with Robust Logic. Humans interacting with an LLM with appropriately chosen prompts can sometimes improve the quality of the resulting reasoning.

Robust Logic offers a specification of reasoning and an explanation of why it is achieved. We have no similar understanding or explanation of the more incidental efforts that LLMs make toward reasoning. It is perhaps little wonder that what these systems get right and what they get wrong look haphazard. Indeed, there is no known justification for following recommendations or adopting beliefs generated by a pure LLM. Systems that provide neither principled reasoning nor the source of the information used, can be dangerous to rely on for educable entities like humans. But the prospects for making systems in which these shortcomings are mitigated are good. The rest of this chapter suggests one approach.

Beyond Efficient Generalization

Current usage of the term "AI" in the popular press is largely synonymous with what supervised learning algorithms, such as LLMs, achieve. I will therefore first discuss the roadblocks that current systems encounter when doing supervised learning in the light of some yet to be realized aspirations that are currently much discussed. I will consider four such aspirations in turn. Each one is a common sense requirement that we humans appear to have some capabilities for but that current computer systems for supervised learning find challenging.

One such aspiration is *lifelong learning,* which here refers to the ability to add to or update learned generalizations without having to be retrained on all the examples one has ever seen in one's life. Many supervised learning algorithms are bad at this, including deep learning. For example, one may have trained a multiple output network, where each output recognizes a different animal species. If one adds a further output for recognizing a new species and continues to train the whole network but only for that new species, then the recognition accuracies for the previously learned species will typically degrade. This is different from humans, who can learn new concepts without having to retrain on old ones.

Another aspiration is resistance to *spurious correlations*. This means that if during training, say, to recognize pictures of elephants, all the positive examples had a dark background and the negative ones a light one, then the trained recognizer should do more on future examples than just predict according to the shading of the background. Doing so would achieve efficient generalization for that data source but would embody a dangerous tendency, since predictions will become inaccurate if the data source undergoes an otherwise inconsequential change in background lighting.

The issue of *adversarial examples* arises very generally in machine learning. In one instance it was shown that it is possible for an adversary to change a few pixels in a natural image of a school bus so that the change is imperceptible to a human, but now the trained classifier thinks the image depicts an ostrich.[13] This problem of adversarial examples can be a security issue in many applications, such as driverless cars.

Finally, there is the general area of *explainability*, the desire that a learned system should not only predict the label of a new example but also offer some explanation of its decision. When computers do medical diagnosis, credit card approvals, or investment recommendations, users report that besides the recommendation itself, some further explanation about why or how the recommendation was made would be helpful. Human nervousness in following a black-box recommendation in

important decisions is warranted. If there is no absolute guarantee that the test examples will really come from the same source
as the training examples, which there rarely is in real applications,
there is good reason to be nervous. Of course, explanations by
themselves do not negate *this* rationale for having anxiety.

All four aspirations listed here are reasonable to have for machines simply because humans exhibit some talent for each of
them. Considerable efforts have been made toward realizing
these aspirations, but progress has been slow. This is not too
surprising. The phenomenon of supervised learning that machine learning algorithms exploit is efficient generalization, as
formulated in PAC learning. PAC learning does *not* promise any
of these other four desirables. All it promises is efficient generalization when test data come from the same distribution as
training data. The difficulties experienced in addressing these
desirables provide evidence that machine learning, as currently
realized, delivers efficient generalization, as promised by the
theory, and little more.

Integrative Learning Technology

To discuss what educability can contribute here, I will draw a
distinction between *end-to-end trained systems*, which I view as
the current standard practice and is captured by PAC learning,
and the more general functionality of *Integrative Learning Systems*,
which are designed to have the broader capabilities realized by
Robust Logic, as described in chapter 6. Educability offers further functionality still beyond this, namely, allowing the constituent classifiers to be programmed explicitly and not only
learned from examples. In this section I will restrict the discussion to integrative learning. The addition of programmed rules
will be discussed in the section that follows.

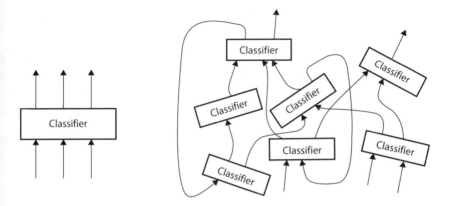

FIGURE 6. *Left*: an end-to-end trained system. The inputs and the outputs are disjoint from each other and fixed. *Right*: an Integrative Learning System. The output of one classifier can be an input of another. One can add further classifiers. The system is integrative if chaining the predictions of the different classifiers constitutes principled inference. An Integrative Learning System becomes an educable system if explicitly described classifiers can be also added.

In current practice, when using supervised learning, say to identify several species of animals, there may be many targets at the same time, but there is just one overall viewpoint, the one determined before the training.

Integrative learning technology would exploit supervised learning in a more flexible way than current end-to-end trained systems. Figure 6 suggests three ways in which integrative learning offers more flexibility. First, the learned system can be composed of several classifiers, trained together or separately. Second, the learning process can uncover previously unknown relationships among the targets. Third, the system has an internal reasoning system that can chain together the predictions of the various classifiers in an *automated and principled* way to reach conclusions that any single classifier may not be able to reach.

Principled chaining is the functionality of integrative learning that was discussed in chapters 6 and 7.

I want to argue now that we should not be surprised about the limitations of an end-to-end trained system, which provides just the single viewpoint envisioned by the trainer. One can obtain some clarity on this simply because one can say exactly what PAC learning promises to deliver, namely, efficient generalization for stable data sources. In the notes for this chapter I give a simple but rigorous proof that the mysterious requirement of generalization is achievable in a certain setting.[14] Accurate generalization is a spectacular enough deliverable. Is it too much to expect more?

Note that in terms of figure 6, a Large Language Model can be viewed as a single classifier, as on the left, but with the outputs feeding in as some of the inputs, as allowed on the right, to allow the text generated next to depend on the text generated earlier.

I shall now comment on how systems incorporating integrative learning have something more to offer with respect to each of the four aspirations described, namely, lifelong learning, resistance to spurious correlations, resistance to adversarial examples, and explainability.

Consider a classifier that makes decisions on whether to approve individuals for a credit card. The features are the components of information fed in about each credit card applicant. The target is the decision whether to approve issuing a card. The company would choose the classifier whose predictions optimize some criterion, such as likely profit. The most profit could come from someone who often pays late, thus accruing interest, but is consistent in repaying their debt eventually, as opposed to people who always pay on time or never at all. The company could use the classifier as a black box, which means that it does not examine the classifier's internal decision-making.

Alternatively, the bank could try to explain to the applicant the reasons for the classifier's decision.

The educability approach would offer more options. In addition to the same features and target, one could put into play further concepts that would be targets and possibly further features. The probabilities that the applicant would get a promotion at work, improve in health, lose their job, or crash their car may all be relevant to the decision at hand. However, an end-to-end trained classifier does not attempt to capture all the knowledge about these real-world eventualities that might be available and permissible to use. The educability framework allows the designer to choose a broader *vocabulary of discourse* that would cover such eventualities that are relevant to the decision but not explicit in the data. The designer would then gather more real-world data that would capture relationships among the concepts in the chosen vocabulary of discourse. This framework enables one to enrich the decision-making in any number of directions. The useful directions would be those in which more information can be gathered and exploited. This framework is more like human decision-making in that it can bring to the task common sense and other relevant background knowledge. In the end, a single decision will still have to be made, but such an Integrative Learning System will be informed in the broader areas specified by the vocabulary of discourse.

How would such an Integrative Learning System compare with an end-to-end training system as far as the four aspirations described earlier?

A child's "why?" is often synonymous with "say more." This is also a useful angle on explainability. An end-to-end learning system would decline an application according to some criterion on salary and number of credit cards already held, perhaps. In contrast, an integrative system would be able to address the request, "Say more about how you judged my case," in terms of a richer vocabulary of discourse that describes real-world eventualities. Humans may find the latter more useful. Explainability is also relevant to professionals, like doctors or investment specialists, when deciding whether to use a computer system as an advisor. Each profession has its own vocabulary of discourse. For instance, an investor may want an explanation of an investment recommendation in terms of concepts of economic theory that are meaningful to that investor. A medical patient would also have a vocabulary of discourse. For example, in a system that predicts an individual's life expectancy, besides receiving the number of years left, the patient might also welcome a broader discourse on what can be done to make that number larger or the quality better. This is exactly what integrative learning has to offer. Note that the systems would make predictions about this broader sets of attributes. There is no need to look inside the individual prediction boxes.

The problem of spurious correlations occurs when a classifier latches on to some feature that is spurious to the real-world classification but happens to work for the available dataset because of a peculiarity of the data collection method. The shading of the background in images of elephants could be one example. This weakness is the result of the single perspective taken in this training scenario. Both a human and an integrative learner would instead have multiple perspectives and have a recognizer for elephant parts—such as legs, trunks, and tusks. In recognizing an elephant, the human or the ILS would check out several criteria, not just one, and make its decision according to the balance of the broader evidence. These multiple perspectives offer some level of robustness against spurious correlations. They also defend against adversarial examples, where the adversary tries to fool the classifier by manipulating the example imperceptibly. The system becomes not foolproof, but a little more difficult to fool.

Last, consider lifelong learning, our ability to add to our knowledge throughout life. We do not need to be retaught every day what we learned last year or in childhood, though reminding helps. It is possible to learn new things and update previous beliefs without degrading the old beliefs. In an ILS, one can learn a classifier for a new target without having to retrain all the classifiers previously trained for old targets. This corresponds to us learning new topics one by one without having to relearn old ones. (Retaining all the examples ever used for training and retraining with all of them each time we see a new example would be prohibitively costly.)

Integrative learning also offers a further generic advantage over single-viewpoint learning: since it takes multiple views, the prediction accuracy of each view can afford to be lower. With this relaxation, more learning algorithms become competitive, including so-called *on-line* algorithms. On-line algorithms examine a sequence of labeled examples, examining each one just once. After each example, the algorithm makes a prediction of the label and is then told the correct label before it moves on to the next example. This is consistent with the intuitive notion of lifelong learning where the human learner may see each example just once.

To summarize, end-to-end trained classifiers have some well-known challenges. These challenges arise from the brittleness of the one-viewpoint end-to-end format and not so much from the specific learning algorithm used. The final arbiter for any safety-critical or otherwise important decision should never be a single such classifier.

Efficient generalization, or PAC learning, is still the power behind prediction, but we must use it more subtly. In integrative learning, one would learn many things about the world, learn how these all relate to each other, predict many things about any one situation, and have principled but automated ways inside the system to chain these predictions. A system for recognizing elephants may, using its many viewpoints, have ten distinct indicators of whether something is an elephant, and its final classifier may decide that any object that has five out of these ten indicators is highly likely to be an elephant. Such a system would conduct a more broadly based evaluation internally and thereby enhance the reliability of its predictions.

Some existing learning systems already consist of multiple modules. For example, the protein structure system AlphaFold, already mentioned, has a module for capturing evolutionary information derived from the protein variants of different species, and separate modules that learn physical folding information.[15] Similarly, software for autonomous vehicles might process information differently from different sensors, like cameras and radar, or about different classes of visual objects, such as traffic lights and lane markings. The integrative learning methodology says that it is possible to have principled designs with many more modules and more flexibility.

One can draw a similar distinction between end-to-end training and integrative learning as I have made between training and education in earlier chapters. Training is from a single viewpoint, foreseen at the time of the training. In education, the goal is to be able to perform in a much broader set of circumstances, including some not foreseen or foreseeable. In particular, in integrative learning the goal is to have a system make good decisions even in circumstances where not enough similar examples have been experienced to train a classifier that would make that decision.

A car on autopilot once crashed into an overturned truck on a highway. The explanation offered was that the autopilot's vision system had not been trained with examples of overturned trucks and therefore had no chance of recognizing such a sight. One way forward would be to train such systems on more situations that are encountered on roads, even if very rarely. This is an inevitable direction to pursue since, as we know, end-to-end or single-viewpoint training can be effective at generalization if one makes a great enough effort. An alternative approach, the one I am advocating here, would be to have an integrative system that is trained from multiple viewpoints and would have figured out that the car was approaching some large stationary object, even if it had *never* been fed closely similar examples in the same context. It would have learned about large objects, moving objects, and stationary objects in other contexts.

A basic way of providing a system multiple viewpoints is to have it sense the world in multiple modalities. We humans have five senses. Cars can be given sensors beyond cameras, for example, radar and lidar, both of which can detect the distance of objects directly. These would have detected the overturned truck. This provides a hint on the nature of the solution I am advocating here. Educable technology would generate multiple viewpoints even on data acquired by just a single sensing device, such as a camera, and would be supportive, clearly, of multiple such devices. Whether the input is from one sensor or more, the system would have separate learned classifiers for as many relevant target concepts as possible. These would include not only the most common objects encountered on roads, but also those that are seldom encountered on roads, for which training sets can be obtained from other settings.

Educable Technology

Integrative learning, as discussed up to now in this section, already has much to contribute to technology. The extra ingredients permitted in an educable system are explicitly programmed rules acquired from the outside and not learned from data. In existing AI systems, such explicit rules sometimes take the form of a *knowledge graph* that captures undisputed facts, such as which pairs of words in English are synonymous and the distance between various cities. It is more challenging for a system to accept from an external source explicit rules that are not indisputably factual. The reason is that it is difficult to verify whether the uncertain rules in the new mix are all valid for the same distribution of situations. Educating such systems therefore comes with the same challenges as educating humans, as discussed elsewhere in this volume. There is no panacea for deciding and providing the explicit knowledge that will be useful for the distribution of cases that the system will encounter in the world.

In spite of these qualifications there is a strong hint that the multiple viewpoint perspective of educability is useful, both with and without explicit instruction. The hint comes from our human abilities. We are not end-to-end trained with a single viewpoint in mind.

At the risk of oversimplification, we can identify three levels of technology that reflect the three stages of evolution that I hypothesize for the development of educability in humans. Level 1 is PAC learning. The cores of current AI systems are essentially this. Level 2 is integrative learning, where level 1 is augmented with some principled reasoning. Here the challenge is to reason in a principled way when multiple pieces of uncertain learned knowledge are being combined. (When uncertain PAC learned knowledge is combined only with certain knowledge, such as certain knowledge graphs, the systems can still be viewed as level 1.) Level 3 is level 2 augmented with explicit instruction so as to achieve educability. Implementing any one level 3 system will come with the same angst as is involved in the design of an education system for humans. There are any number of ways of choosing the teaching materials used in the instruction, and it is difficult, without testing, to predict how effective the application of that knowledge will be in the world in which the system is to function.

Natural versus Artificial

An ever-present question in engineering is the extent to which one should seek to imitate nature when designing artificial devices. As previously discussed, in choosing which *tasks* computing devices should perform, humans have up till now derived most of the inspiration from human capabilities and desires. The open question has been the extent to which we should also imitate nature's *algorithms* and *methods*.

The context of educability raises the question of whether in this area we should consider *not* emulating nature's task. An animal species is composed of a large number of individuals with short life spans compared to the longevity of the species as a whole. Each individual can learn from its experience and transfer its information to others only during its short life. Before humans had appeared, the information that a single individual could transfer to another was limited. The solution evolution found that enabled an individual human to amass a large amount of world-tested knowledge was to improve the ability of individuals to transfer knowledge among one another.

Is this formulation, a population of individuals accumulating their knowledge through communication, only relevant to short-lived organisms? If one has a single machine of arbitrarily large memory capacity, then is the architecture of educability, based as it is on individuals learning separately and then communicating with each other, still necessary?

I suggest that the answer is *yes* for reasons that arise both from the properties of the world and from those of foreseeable technologies. As far as the modularity of experience that the world offers, there may be little difference between humans and machines when they face similar experiences in the world. As far as the modularity of technology, current artificial devices are modular just like a biological species is modular, in being composed of independent processors just as a species is composed of independent organisms. Each module in a large digital system has its own resources, but only a limited amount when compared with the entirety of the system. In this it is similar to an individual animal in the context of its species. Locality in space is important for a computer module for optimizing efficiency since short-distance communication is most efficient. Space localities also arise from the fact that certain data is available only at certain places. For these reasons, I would argue that the educability model has relevance beyond short-lived organisms.

The basic question is the one of *maximality*: Is educability the most powerful model of its kind, setting aside that its numerical parameters can take arbitrary large values? If so, then the intellectual nature of machines will asymptote, at least qualitatively, to the human capability of educability and no more. This question is of some importance. If the answer is yes, as I suspect, then we humans can expect to coexist with machines in a mixed economy, where some intellectual services are provided by humans and others by machines. Currently there is human demand for playing chess against both humans and machines, although one might have predicted that this would not be the case when machines are so much better than humans. Similarly, stock market investment decisions are shared between humans and computers, and checkout at many supermarkets also offers the two options. All this involves human preferences, which are difficult to predict. Some humans may decide that they only want to watch movies acted by avatars and written by computers in the style of Shakespeare. But I suspect that in the future there will continue to be a mixed economy in intellectual services provided by humans and machines. It will be difficult for machines to outflank the diversity and ever-changing nature of human want.

The Challenge of Teaching Materials

Humans have invested enormous effort in developing teaching materials for educating the young. Customarily, instruction through secondary school and some way beyond uses material that has been carefully curated, many times revised, and widely tested over a long period. The overall results have been spectacular. With our biological hardware and learning algorithms seemingly unchanged for thousands of years, the enormous increase in our cognitive productivity must derive in part from what is contained in these teaching materials.

For machines, whether using basic machine learning or the methods advocated here, the choice of teaching materials will be equally important. As mentioned, some datasets, such as ImageNet, have already been highly influential in the development of machine learning. While advances in learning algorithms are critical also, they will be of little use if appropriate teaching materials—carefully prepared and curated—are not available to supply the content.

The preparation of useful teaching materials has its challenges. There may be plenty of data on the web, but how does one tell what is factual from what is fictitious?

Challenges already arise for factual data. Consider a basic machine learning system with a small overall rate of error in generalization. If the data concerns a human population of which 10 percent is a minority group, then whether doing face recognition or medical diagnosis, the standard method of sampling individuals equally will result in a factor of nine fewer examples of the minority. If the minority differs from the majority in the property to be predicted, then the predictions will be less accurate for the minority.

For this and other reasons, one can expect that the creation of educable technology will require a constant effort to develop new datasets or teaching materials to fit the evolving requirements. This would complement the enormous efforts constantly made to keep up with the steady expansion of knowledge when teaching humans. The challenge of continuously creating new teaching materials for machines will persist, I expect, long after the current questions about learning and reasoning algorithms are largely settled.

This will maintain the practice of AI as a highly humanistic endeavor. Think of current Large Language Models: they are trained on texts written laboriously over the years by possibly hundreds of thousands of human authors and then further trained by a large number of present-day humans, all with the aim of pleasing humans.

No Technological Singularity

The question often raised is what will happen when machines become more "intelligent" than humans. That hypothesized moment has been called the *singularity*. The question is often accompanied by the suggestion that there is much to be feared. In the last decade many saw computers recognize their voice or face, or make apparent conversation, and thought immediately that the world had changed forever and the long-feared singularity when machines would take over was at hand. Many have expressed alarm.

Human societies have experienced loss of control when invaded by other societies. The inhabitants of sacked cities in the ancient Middle East and many of the Indigenous peoples of the Americas experienced such loss. These events were singularities for those societies. The hypothesized "AI singularity" would be different in not being the result of external forces. It would be a consequence of factors currently well within our control.

The following argument from a paper by I. J. Good, Turing's sometime collaborator, in 1965 is widely quoted.

> Let an ultraintelligent machine be defined as a machine that can far surpass all the intellectual activities of any man however clever. Since the design of machines is one of these intellectual activities, an ultraintelligent machine could design even better machines; there would then unquestionably be an "intelligence explosion," and the intelligence of man would be left far behind.[16]

Does this argument doom humanity, or is it flawed? The argument assumes that intelligence comes in arbitrarily powerful varieties. I believe this assumption is dubious. Not everything comes in arbitrarily powerful varieties. For example, as we have seen, it is generally accepted that the power of computation, aside from efficiency, is already maximal with Turing computation. Also, we should not make assumptions about concepts such as intelligence that we cannot even define.

Educability offers a vocabulary for discussing the issues raised by Good's argument. In the educability model, there are several numerical parameters, such as memory size. When comparing two systems, one system may be better in one parameter and worse in another. But the larger these parameters are, the more effectively the systems can receive education and apply it. There can be machines that exceed the parameters of humans in every dimension. Even then, however, the performance will come from the quality of the education the machines have received as much as from the raw computing power of the machines themselves. Bigger and bigger machines that are well-educated and replete with knowledge relevant to the task at hand will exceed the performance of humans in every sense that can be defined in these terms. Setting them against ourselves in competitions we are sure to lose does not make sense, and surely we will strive not to do that.

AI—a phrase that in this section I use in the currently conventional sense—does indeed carry dangers. There are grave dangers to developing weapons with advanced capabilities of perception and reasoning. As with other dangerous technologies, international agreements to control them will need to be crafted. In a recent book, Kissinger, Schmidt, and Huttenlocher argue that military and cyber applications of AI present problems for the international order that will be challenging to manage.[17]

The singularity fear is sometimes associated with the assumption that a highly intelligent machine will necessarily acquire the biological characteristic of ruthlessly acting in its own interests, at the expense of humans. I think this fear is misplaced. Clearly, we should not deliberately manufacture machines with these ruthless characteristics. Nor should we let loose an evolutionary mechanism that breeds machines with these characteristics. I do not believe that AI is some uncontrollable force that will overwhelm humanity by its very nature. Computer technology is powerful, and catastrophic things may happen by accident or through human greed or malice, but they do not need to happen. We need to prevent such negative outcomes as a society in the same way that we control the possible negative consequences of other dangerous technologies, such as nuclear weapons or gasoline-powered cars.

The main safeguard we have is science. If AI were a mysterious force that is incomprehensible to us, then it might be truly dangerous and uncontrollable. I believe that we will continue to understand AI in terms of capabilities that we can define. Educability is one such framework. Our protection will be that we will only build systems that are based on capabilities supported by known science.

I believe we will have sufficient understanding of the science behind AI as it advances to make reasonable judgments of the behavior of systems that are based on that science. We will seek to control their deployment to the same extent that we seek to control our many other dangerous technologies. We have reason to be optimistic simply because we have an understanding of what its scientific basis is. Such an understanding does not mean that we can predict exactly how an AI system will behave in all situations. Learning algorithms have many moving parts. For most practical learning algorithms, we do not know how to predict their exact behavior on a new dataset. The same holds true for many other applications of computers where practical algorithms often have good behavior, though we cannot predict exactly when. Nevertheless, we do have methods for determining what these systems achieve. For example, we can check how accurately a learning algorithm predicts on a dataset by training it on part of the dataset and then testing it on the rest.

The dangers of AI are malicious or reckless misuse, accidents and misjudgments, not its inherent nature. As with other powerful technologies, we need to invest in understanding relevant safety issues. We will need to anticipate likely sources of accidents and likely methods of malicious misuse. The reason for being optimistic about managing these dangers is the expectation that we will continue to understand the science behind the practice well enough. AI safety will not be possible without the science.

One recommendation for AI usage that current science strongly suggests, I believe, is that all products of AI systems should be labeled as such. This should apply to products in all media, whether text, pictures, videos, or sound, and even if there is no deepfake intent in the production. The reason is that current systems at their core are "probably approximately correct." Such outputs may be suitable for entertainment, or to generate alternatives for a human to subsequently choose from, or for achieving a statistical edge when applied many times as, possibly, in investing. These outputs may even have higher accuracy than humans on a recognized dataset. But for now, humans need to know. Our acceptance of information we cannot validate for ourselves needs to be based on trust. We need to know, just as we have always had to know, who or what is providing the information.

But I do not think we need to panic about AI. There are many shouting "Fire!" This can generate both useful discussions and dangerous behavior. Unqualified warnings have the risk of exaggerating the power of existing AI systems, giving them the aura of the mystical and incomprehensible and thereby inviting misuse. AI as currently realized is the result of decades of scientific research that has yielded steadily growing capabilities *and* understanding. It is possible and necessary to have informed discussions about AI.

There is a long history of deployment of new technologies and society having to play catch-up in dealing with the unintended side effects. The long-term effects of the motor car on how cities evolved and on how we live are still playing out.

Rules to regulate road traffic evolved with the growing traffic. How best to power vehicles without causing climate change is a continuing concern. In the present century the deployment of digital technology is having similar large-scale side effects on how we live. It is right that we should be thinking ahead, anticipating the likely consequences as much as we can, and heading off ill effects. This technology is the latest spectacular outcome of our species' educability. To enjoy its benefits, we have to pay a price in caution, but I know of no reason for panic.

We should not be fearful of a technological singularity that would make us powerless against AI systems if we cannot define any sense in which they would be decisively different in kind from us: as far as we currently can tell, the best they would be able to do is to be educable like us. A singularity has already occurred on Earth. It was the one that made many non-human species powerless relative to humans. It occurred as a result of the biological evolution of educability in humans.

CHAPTER 12

Education

SOME QUESTIONS

Education as a Right

Education is the object of much lip service. Its importance is frequently asserted. That education is a human right is enshrined in the Universal Declaration of Human Rights adopted by the United Nations General Assembly in 1948. Article 26 says the following:

1. Everyone has the right to education. Education shall be free, at least in the elementary and fundamental stages. Elementary education shall be compulsory. Technical and professional education shall be made generally available and higher education shall be equally accessible to all on the basis of merit.
2. Education shall be directed to the full development of the human personality and to the strengthening of respect for human rights and fundamental freedoms. It shall promote understanding, tolerance and friendship among all nations, racial or religious groups, and shall

further the activities of the United Nations for the
maintenance of peace.
3. Parents have a prior right to choose the kind of education
that shall be given to their children.

The educability notion provides perhaps the simplest explanation of why education is a human right: humans are educable
in the same sense as fish swim and birds fly.

An operational question is whether understanding educability will help to better realize and deliver this right. The text of
the declaration reminds us that some central issues still need
clarification.

For instance, the second paragraph acknowledges that instill-
ing beliefs, such as "respect for human rights and fundamental
freedoms," is within the power of education and that education can be a means to promote certain beliefs, such as "tolerance and friendship . . . maintenance of peace." However, the
question of how exactly education influences beliefs is not
well understood. It is a worthy subject of study and one that,
I hope, the notion of educability will further encourage. Students in the United States learn to recite, from the Declaration
of Independence, that "all men are created equal." What do they
understand this to mean? Do they act on this belief as they understand it?

The Study of Education

Immense as the resources invested in education may be around
the world, we still do not give the subject as much respect as it
merits. We may be giving it our hearts, passions, and sweat, but
not enough of our brains. If educability is our species' defining
trait, then education should be its central concern.

One would expect that universities would offer their under-graduates the chance to devote themselves to the most funda-mental and important fields of inquiry of their time. At Harvard, the institution with which I happen to have most familiarity, education is offered to undergraduates only as a secondary field of study, and not as a major. The situation at many similar insti-tutions in the United States is about the same.

There is some irony here. A search through the Course Cata-log for the Faculty of Arts and Sciences of Harvard University reveals the myriad facets of education that are the subject of individual courses. The titles offered in the various departments include The Politics of Education in the Developing World; The History of African American Education; Philosophy of Educa-tion; Educating Incarcerated Youth; Sociology of Higher Edu-cation; Appraising and Reimagining Middle and High School Math Education; Educational Justice; Growth, Technology, Inequality, and Education; The Role of Music in Health and Education; Higher Education: Students, Institutions, and Con-troversies; Educational Outcomes in Cross-National and Cross-Cultural Perspectives; Education in the Economy; Transforming Tradition: Islamic Education in the Modern Muslim World; Student Leadership and Service in Higher Edu-cation; Democratic Citizenship and Education; Democracy and Education in America.[1]

The number and variety of these topics give evidence of the importance of education to so many aspects of human life. These courses, however, are not asking the most basic questions about education. They do not purport to. Many are designed from the viewpoint of another field and offered by a depart-ment with a mission other than education. A Harvard under-graduate could be asking other questions. What is the power of education? What are four years here going to do to me or my brain? Most of the courses listed above address these questions only tangentially, if at all.

Is there something puzzling about this state of affairs? While the *delivery* of education attracts gargantuan resources, the *understanding* of education is losing out to scores and scores of other fields. Ironically, it is the institutions of higher education that are making this choice.

A Different Vantage Point

Not all questions have a science-based answer. Here is one argument that suggests that the question of the nature of education does have one: Darwinian evolution of species and learning from experience by individuals are both natural processes that occur on a vast scale on this planet. These are the two main ways in which information has flowed into living organisms over the millions of years. We expect these processes to be subject to scientific understanding. Education is a third method by which information now flows into living organisms, namely, humans, and again on a vast scale. We would expect that it should be similarly amenable to scientific understanding.

The idea that education has an underlying science has been asserted many times. In 1806 Johann Friedrich Herbart published an influential book *Erziehungswissenschaft,* or *Science of Education.* The phrase "science of education" has been often used since and into the present day. Enormous volumes of empirical research findings are available.

Educability, as defined in this book, starts from a basic science perspective. It identifies three essential pillars and combines them. These three pillars are (a) learning from experience, (b) acquiring theories through instruction, and (c) applying what one has acquired through (a) and (b) in an integrated way.

Since antiquity, diverse philosophies of education have been proposed, and varied institutions have been founded based on them. While the philosophies and institutions are too numerous to list, many of them can be related to some of the three pillars. In the past few centuries, a common strawman characterization of traditional education has been that of a teacher reciting theories that the student needs to memorize. This "drill and kill" approach is a caricature of exclusive emphasis on pillar (b)—theories transferred by instruction. This characterization is often contrasted with various kinds of reformed education, such as Montessori, where the student is more active and learns in a more experiential or hands-on manner, which brings in pillar (a). One of the earliest recorded educational theories is the Socratic method, recorded by Plato, which emphasizes logical thought in the context of beliefs and involves pillar (c).

I am suggesting that in the human cognitive apparatus that pertains to educability, all three pillars—(a), (b), and (c)—are fundamental. The corollary of this is that, to exercise all parts of this apparatus, we need educational offerings that combine all three.

This volume points to a new vantage point from which to understand the possibilities and limitations of education. If our capacity for education *is* the power behind our current science-based civilization, it would be the ultimate irony if we failed to pursue every possible angle toward a science-based understanding of this power.

In the Classroom

In certain very particular areas, such as the teaching of reading, concerted research efforts have been impactful in influencing national guideline in some countries. More generally, however, as many have lamented, existing research on education has had less impact on the practice of education in the classroom than hoped for. One would expect that research would have more impact if it were recognized to have more of a scientific basis, if it succeeded in addressing education at a more foundational level and if it gave rise to a greater consensus on some core tenets.

There is a large literature on educational psychology.[2] Extensive empirical studies have been done to compare different pedagogical styles for effectiveness in practice.[3] There have also been many attempts to organize existing knowledge about this science of education as courses and textbooks. An informative example is the text *The ABCs of How We Learn: 26 Scientifically Proven Approaches, How They Work, and When to Use Them,* by D. L. Schwartz, J. M. Tsang and K. P. Blair. It is based on a Stanford University course, "The Core Mechanics of Learning", described as "an applied course that emphasizes learning theories that can be put into practice." As the title suggests, the subject is organized as twenty-six topics, one for each letter of the alphabet. This organization is appropriate to the current state of knowledge about education, I believe, but betrays the best practices orientation of the field.

The much-used term "learning outcomes" refers to statements made at the beginning of a course describing what students should know, be able to do, or value as a result of taking the course. There is much current encouragement for instructors to compose such statements. This is mostly to the good. It is an attempt to define what a course seeks to achieve, analogous to how one wants to define what education achieves. Nonetheless, it does have a paradoxical nature. As I have said before, for training, one knows exactly what one wants the result of training to be. Hence for courses of training, learning outcomes are entirely appropriate. But education is different because one wants the result to be broad enough to encompass the unforeseen and unforeseeable. As Oscar Wilde put it: "Education is an admirable thing, but it is well to remember from time to time that nothing that is worth knowing can be taught."[4]

The approach taken in this book is to propose a concrete definition of educability in terms of computational processes. These processes encompass how information is acquired and how it is applied. It is intended as a description of the core processes of education. For understanding education to the full as it relates to humans, complementary approaches are equally essential. For many decades, psychologists, animal behavior researchers, sociologists, neuroscientists, and others have been using their various methodologies for approaching issues that are close to those I have discussed. The methods will need to be used in tandem if we are to better understand education and improve its practice.

I have been suggesting that we humans have clearly defined capabilities that enable us to receive and apply education on a massive scale: learning by example, receiving instruction, and chaining. Some parameters, such as the belief choice policy discussed in chapter 8, qualify these capabilities. Psychological experiments can explore these parameters systematically. All this provides a focused scientific question: Given that humans have the capabilities that we do, what is the best way of taking advantage of these capabilities to improve education?

I make a distinction between the basic *capability* of educability and the process of *harnessing* this capability. A major concern of educators is that students of similar apparent potential often progress through the education system with different levels of success. These students may be going to different school systems, or to different classes in the same system, or to the same classes. With a better understanding of educability, could we better harness the capabilities of all students so that society would benefit more from their talents?

How does one motivate a student to invest effort in a certain subject? What is the role of grading and other rewards in motivating students? How does classroom culture influence the motivation of students? Empirical findings on the best way of harnessing educability are, of course, valuable. Much of the educational literature referred to above, as this author reads it, focuses on that, as opposed to the nature of the capability that is being harnessed.

A widely held belief about education is that it is inherently good. The more education an individual receives, and a country offers, the better. Education will automatically cure the evils of the world. History offers some cautionary warnings.

After World War II it was widely observed that things are not quite this simple. Germany had enjoyed as much education as any other country, and more than most, and yet it had unleashed some of the greatest evils known to history. World War II had punctured the nineteenth and early twentieth centuries' high hopes that having more educated populations would bring about an age of never-ending progress. This historical record provides evidence that we do not understand education or its effects too well.

Despite this, I do believe that education remains humanity's best hope. To find our way, we need a more nuanced understanding of what exactly education is and can provide, and how it influences an individual and a society. I am suggesting that one can investigate these questions from a more scientific perspective than hitherto. The various Harvard courses referred to earlier relate mostly to existing or past education systems. They relate much less to the nature and potential of education itself or the impact of its various possibilities on individuals and societies. The study of education as an academic discipline does not have the highest status at present. I am suggesting that it can, if pursued with the focus it deserves.

One can apply educability theory to educators. The hardworking people currently employed in education may be using their abilities to the full to learn from their professional experience and to apply what they have learned in their work. However, if there has been an underinvestment in the understanding of education, then educators would not be fully utilizing their abilities to themselves absorb instruction. The emergence of a widely supported more foundational basis for education could serve to better support teachers and increase the confidence that parents and society have in them.

Teaching Materials

To be serious about education, we need to be serious about the content that is taught. The choice of materials to include is ever more challenging because of the ever-increasing accumulation of recorded knowledge. Societies need widely supported methods of agreeing on their school curricula.

H. G. Wells in the 1930s provided what he regarded as an urgent rationale in response to his times, when governments with mutually inconsistent ideologies—fascism, communism, and liberal democracy—were in control of the most powerful nations. He promoted the idea of a "World Encyclopaedia" that would contain carefully compiled knowledge that experts of the time could agree on. He claimed that "without a World Encyclopaedia to hold men's minds together in something like a common interpretation of reality, there is no hope whatever of anything but an accidental and transitory alleviation of any of our world troubles."[5]

Encyclopedias aim to bring together teaching materials systematically in a way that is authoritative, organized, and comprehensive. The word encyclopedia comes from the Greek *enkyklios* (ἐγκύκλιος), meaning "general," and *paideia* (παιδεία), meaning "education." Encyclopedias go back to at least Roman times. Pliny the Elder, living in the first century, claimed to have included in his *Naturalis Historiae* 20,000 facts taken from 2,000 separate works by over 200 authors. In the Renaissance, there was much awareness of the importance of preserving knowledge, given that much from antiquity had been lost. The French Enlightenment era *Encyclopédie*, published from 1751 onward, had 28 volumes, with 71,818 articles. Edited by Denis Diderot and Jean le Rond d'Alembert, it had over 150 contributors. It met with political opposition from the French government, reflecting the challenge of giving authoritative accounts of controversial matters.

Can or should the world be held to a "common interpretation of reality"? Clearly, excessive divergence of interpretations can provoke unnecessary conflict. To minimize that, a reasonable goal for a nation may be to have more mechanisms for producing authoritative accounts, accepted to be politically unbiased, of the available facts about a question of topical interest. An example of such a mechanism is the US Congressional Budget Office, which conducts nonpartisan analysis of the budgetary consequences of proposed legislation.

Education and Propaganda

As Nelson Mandela said, "Education is the most powerful weapon we can use to change the world."[6] It is hardly surprising, then, that it can be used for nefarious purposes. Josef Stalin had already told us that "education is a weapon whose effects depend on who holds it in his hands and at whom it is aimed."[7]

A price we pay for the benefits of educability is our fallibility in judgment when absorbing and committing to new theories. Political propagandists and public relations practitioners are ever ready to exploit this human frailty. Individuals may be more or less committed to their beliefs. The goal of propagandists is to modify these beliefs using the limited interventions available to them.

Forming and manipulating belief systems are both phenomena of educability. In the discussion of belief choice in chapter 8, we saw that young children already have methods for deciding which of several competing opinions to believe. Psychologists have also explored the question of how adults choose between competing theories. People in public relations and political campaigning have been paying attention.

There are widely studied psychological phenomena that concern our readiness to accept new beliefs in terms of the overall consistency of our belief system. These studies explore how our current beliefs influence additional beliefs we might be inclined to adopt. The principle of *cognitive dissonance*, formulated by Leon Festinger, asserts that humans are uncomfortable with actions or beliefs that are contradictory and will seek to modify their actions or beliefs to reduce such contradictions.[8]

On certain questions, such as evolution, climate change, or vaccinations, the scientific community has reached near consensus, but the general population holds a wide range of beliefs. This phenomenon offers many opportunities for studying how individuals arrive at their beliefs, or in our terminology, their belief choice. One idea is that humans have a strong tendency to adopt beliefs that they believe other members of the group they most identify with hold. A related theory is that of cultural cognition, which asserts that political beliefs bias judgment of risk versus reward. Another idea posits that one source of resistance to widely accepted scientific theories is interference from more intuitive theories already held.[9]

These belief choice policies determine the influence of existing beliefs on our willingness to adopt a new theory. Other biases concern who is offering us the new theory. An influential book in this area was *Personal Influence* by Katz and Lazarsfeld, published in 1955. They discussed the role of *opinion leaders* in influencing public opinion in the context of mass media. Using opinion leaders to influence the public is now a widely adopted technique.

Are there other ways for propagandists to influence our behavior? One angle has been the idea that the behavior of crowds is different from that of individuals. A source for this was the book *The Crowd: A Study of the Popular Mind* by Gustave Le Bon, published in 1895 in France.[10] It asserted: "Before we can designate a group of people as 'a crowd' or 'a mass,' the group must have undergone a transformation where its members are guided by a collective emotionality and uncontrolled instincts, which in turn makes it homogeneous. As a consequence, the personal sense is for a moment overthrown by a collective ecstatic state of mind." The fear articulated there was that, with the rise of democracy, crowd behavior would somehow destroy civilization by displacing rational thought. This theory that crowd or herd behavior is different from individual behavior and more easily manipulated was influential on twentieth-century dictators. A different phenomenon is the so-called *risky shift* principle, which is related to *group polarization* and asserts that groups are willing to take riskier actions and hold more extreme positions than the constituent members would individually.[11]

There is also psychological research that shows how the provenance of rules and their consistency with our current beliefs interact with each other in our belief choice. If instructed to do so by an authority figure, humans are surprisingly willing to perform actions that, on the surface, are diametrically opposed to their principles. Most famously, in the 1960s Stanley Milgram had human subjects believe that they were performing word-association tests on some learners, who were Milgram's confederates. He instructed the subjects to administer electric shocks whenever the learner made a mistake, with the level of shock increasing each time. When the learners pleaded that the experiment be stopped as the shocks were getting too painful, the subjects had to decide whether to stop or continue following instructions. A majority of Milgram's subjects carried on, administering what they thought were more and more painful levels of electric shock.[12]

Understanding belief choice in humans is complicated by the fact that different individuals may be using different concepts when making decisions about the same question. The area of political choice makes this clear. How do individuals choose whom to support at elections? One study divided up the US electorate of 1956 into five groups according to the following categories: (1) Ideologues, (2) Near-ideologues, (3) Group interest, (4) Nature of the times, and (5) No issue content.[13] The first category consisted of "those respondents who did indeed rely in some active way on a relatively abstract and far-reaching conceptual dimension as a yardstick against which political objects and their shifting policy significance over time were evaluated." The second were those "who mentioned such a dimension in a peripheral way but did not appear to place much evaluative dependence upon it or who used such concepts in a fashion that raised doubt about the breadth of their understanding of the meaning of the term." The third category "evaluated parties and candidates in terms of their expected favorable or unfavorable treatment of different social groupings in the population." The remaining groups, like the third, appeared to make no ideological judgments. The fourth included some who judged by a "single narrow policy for which they felt personal gratitude or indignation" without seeing that as "representative of the broader policy postures of the parties." The fifth included those "who felt loyal to one party or the other but confessed that they had no idea what the party stood for" and those who "devoted their attention to personal qualities of the candidates, indicating disinterest in parties more generally."

This study estimated that of the voting population, only 2.5 percent were Ideologues and 12 percent Near-ideologues, suggesting that the theories that different individuals use to make the same political choice vary widely. Only these first two groups, who accounted for less than 15 percent of the population, used political theory, in the sense of left versus right, or liberal versus conservative. This phenomenon has not been lost on propagandists, who know that they do not need to change basic political leanings in this theoretical sense. It is sufficient to influence in other ways. The propagandist will be suggesting that if you vote for X, you will somehow benefit, that X is like you, or that X has good personal qualities. In recognizing whether you will benefit or whether someone is like you, you may be invoking your existing beliefs, but these beliefs need not be about political theory.

Propaganda Safety

The question of how to harden individuals to be more cautious about incorporating new material into their belief systems is one that is topical and challenging. Propagandists who aim to modify the belief systems of large groups abound and have any number of new ways to operate through electronic media. One imagines that since prehistoric times there have always been

individuals who have had an interest in promulgating a particular belief system. However, the current frantic competition for
mindshare may be unprecedented.

The starting point of the educability hypothesis is that the
basis of cognition is efficient generalization and its elaboration
to integrative learning. Both these concepts are learning neutral, in that they are defined in the first instance for a world that
is trying neither to help nor hinder the learner. Educability is
an adaptation of this basis that additionally provides the ability
to be instructed. This adaptation is *learning enhancing* in that it
gives a role to an external agent, the teacher. As mentioned previously, nowhere in this system is there anything that copes
with *learning adversarial* situations, where an external adversary
is seeking to disrupt the learning process to further the adversary's agenda and perhaps harm the learner.

This leaves human nature naïve and gullible. Woe be to the vul-
nerable human in the early twenty-first century. Public relations
has been a thriving industry for more than a century, designed
to influence the opinions of the general population. Billions are
spent each year in its practice for product advertising, without
much public debate. Political propaganda has existed since time
immemorial. Rhetoric, the art of speaking to influence opinions, was already highly developed in ancient Greece. At every
election, in every country, its techniques are deployed to influence
voters, again without much public discussion of the techniques
themselves.

With the proliferation of electronic media, these techniques
can be employed on a greater scale and with more accurate
targeting than ever before. Legal guardrails against one-sided
presentations are few. For example, in the United States the "fairness doctrine" of the Federal Communications Commission
required broadcasters from 1949 onward to present controversial

issues of public importance in a balanced way. This doctrine was eliminated in 1987.

Saying simply that individuals should be more rational or more fact-based is underestimating the difficulty. There is a substantial challenge in understanding these human frailties and in mitigating their consequences. Evidently, we ought to evaluate our belief systems against facts more carefully than we routinely do. For this, we need to evaluate our sources of information for reliability more carefully than we do, especially when we know that the matter under consideration is controversial. How to do this exactly is not so clear.

When the Buddha was asked how to tell among competing philosophies that all purport to tell the truth, he replied as follows.

> Please, Kālāmas, don't go by oral transmission, don't go by lineage, don't go by testament, don't go by canonical authority, don't rely on logic, don't rely on inference, don't go by reasoned contemplation, don't go by the acceptance of a view after consideration, don't go by the appearance of competence, and don't think "The ascetic is our respected teacher." But when you know for yourselves: "These things are unskillful, blameworthy, criticized by sensible people, and when you undertake them, they lead to harm and suffering," then you should give them up. . . . But when you know for yourselves: "These things are skillful, blameless, praised by sensible people, and when you undertake them, they lead to welfare and happiness," then you should acquire them and keep them.[14]

Does educability offer any solutions? As discussed in chapter 9, an Educable Learning System can be taught to reason in multiple ways. While the basic mechanism is gullible, one can teach it ways of reasoning that would prepare it to avoid some pitfalls. The Buddha was describing a particular—though highly sophisticated—strategy for belief choice.

Here is a concrete proposal for furthering *propaganda safety* by means of education. Schools should educate children about how propaganda is, can be, and has been used for manipulating opinions. The population should be made knowledgeable about the methods by which false information can be spread by individuals, by print, and by digital media. Everyone should know about recently developed methods of faking digital information. Everyone should be alerted to, educated about, and hardened against becoming victims of propaganda and false information. It is important to protect ourselves from the new dangers made possible by deepfake and personalized messaging technology.

The more fundamental and ancient problem, however, is our own vulnerability as humans to being misled even by old methods. Telling fact from fiction can be difficult. Writers from time immemorial have been skilled at confusing us. When watching a movie or reading a novel, telling whether it is based on fact often seems impossible. Financial fraud continues to flourish. We need to be educated to be mindful of how easily we can be fooled, even by traditional means.

Nuclear safety and AI safety may be manageable under normal human conditions because of their science base. They are most likely to suffer failure if some group or nation is persuaded through propaganda to pursue a reckless course. In other words, nuclear safety and AI safety are most likely to fail only after propaganda safety has failed. Pursuing propaganda safety via education might make H. G. Wells's observation that "human history becomes more and more a race between education and catastrophe" concretely prescient.[15]

Saturation

I started this volume by suggesting that our species may have had its special capability, educability, from the beginning more than three hundred thousand years ago, but the capability may have languished, seriously underutilized, for much of that period. Here I am suggesting that we may be living in the first century in our history in which we as a species are in a position to fully utilize this capability. At the fingertips of anyone with a web-enabled cell phone is the possibility of spending all day educating themselves with college-level courses, or reading fiction, or consuming political commentary. The rate at which we ingest this material is limited only by our ability to absorb it. In this sense we have reached saturation point. Only a few thousand years ago, before the invention of writing, access to knowledge was severely limited for everybody. Now, for those with unrestricted online access, our species' ability to absorb knowledge may be fully saturated by the availability of knowledge. The saturation point means only that we are able to access knowledge at the maximum rate that we can absorb. It does not guarantee that we will use this opportunity well.

The earlier technology of printed books has been preparing us for meeting this challenge by providing expanding access to knowledge at a steady pace. Universal education was the mechanism by which much of this broadening access was realized. Some dry statistics provide a view of our progress in this. A summary of the state of current efforts is available in *Education at a Glance 2022: OECD Indicators,* published by the Organisation for Economic Co-operation and Development. It refers to the thirty-eight OECD member nations with developed economies, which between them account for more than 60 percent of the world's nominal Gross Domestic Product. One statistic, on the positive side, is that in 2021 the mean percentage of twenty-five- to thirty-four-year-olds who had tertiary (i.e., college) education was 47.5 percent, with five nations (Canada, Ireland, Japan, Luxembourg, and South Korea) exceeding 60 percent. Most encouragingly, these numbers are rapidly increasing. The OECD average was only 26.8 percent in 2000.[16] The statistics suggest that further expansion in the delivery of education in the next decades can be expected.

There remains much room for improvement. In 2019 OECD average expenditure of national wealth for primary, secondary, and tertiary education was only 4.9 percent of GDP.[17] Also, the report quotes average salaries for lower secondary teachers as a proportion of the salaries of similarly educated workers in their countries. This proportion was 0.9 as the average for OECD members, and 0.6, for the United States.[18] If education is as central to our species as this volume suggests, then teachers are currently underpaid.

If educability is the critical human trait, then we should see education, in all its forms, as the central human enterprise. A human life is one in a continuous state of learning and changing. We are often reminded that in the current world we will need to retrain for several different careers. The educability notion offers the good news that as a species we are well equipped for this challenge. For the first time a substantial fraction of the population may be living up to the human potential of doing work that is challengingly new every few years.

CHAPTER 13

A Species Adrift

Human Sacrifice

Human thirst for belief systems is endless. We devour fiction, fantasy, myth, and other narratives, absorbing tortuous details about individuals and worlds that have no relevance to us or may not even exist.

One apparent byproduct of this proclivity is the practice in times past of human sacrifice—the ritualistic murder of our own species to please a deity. Murder of one's own species is not unique to humans, but the drive to do that from an imagined abstraction does appear to have our signature.

Human sacrifice was not a mere oddity but a pervasive characteristic of our history as a species. Many of the practices continued into recent centuries and have left records. For others, there is ample archaeological evidence. The Aztec apparently believed that for the universe to continue one needed to appease the gods by the sacrifice of thousands. Archaeological evidence shows that in Peru the Inca sacrificed hundreds of children in a single day. Human sacrifice was not localized to the Americas. At various times it was widely practiced also in Europe, the Middle East, China, India, Africa, and the Pacific

Islands.[1] In one study of ninety-three Austronesian traditional cultures on islands in Southeast Asia and the Pacific, it was found that forty among them had practiced human sacrifice prior to recent contact with outsiders.[2]

Societies can be taken over by belief systems that their descendants find unconscionable and incomprehensible. Presumably, significant fractions of the relevant populations, including their best-educated members, accepted the beliefs that occasioned these occurrences of ritual slaughter. In each case, our ancestors were persuaded of some theory that normalized the murder of their fellow human beings. Such theories may have varied in their particulars, and many have vanished into history, leaving little trace.

I am suggesting that the characteristic that makes humans different from other species, in this respect of ritual slaughter, is none other than the one that enabled civilization: our educability—our facility with absorbing arbitrary belief systems. We have an endless facility not only for absorbing belief systems, but also for acting on them. We pursue a multitude of trades and professions, most requiring an inordinate amount of specialized knowledge. No other species comes close in the granularity of this division of knowledge and labor. We also have varied political belief systems that are inconsistent with one another but often held with total conviction. Religious belief systems are equally strongly held and similarly diverse. Interpretations of national histories and cultures have multiplied without end, as have views of literature and music. Charles de Gaulle once asked: "How can anyone govern a nation that has 246 different kinds of cheese?"

We do not understand why human sacrifice was once so pervasive, or why it has largely vanished. The surprise is the ease with which societies can adopt belief systems that later become untenable.

The viewpoint I have been presenting is that the acquisition and processing of belief systems is a computational process that does not judge the beliefs themselves. The basic processing treats all belief systems as equal. On the other hand, we individually do judge beliefs as being different—some as good and others as bad. Do we need to be concerned that there seems to be no protection that the ruling belief systems of the future will not be more detrimental to us than the current ones? I believe there is reason for concern.

At Sea

I have been suggesting that the computational capabilities that had evolved in mammals developed further in our more recent ancestors until they finally crossed the threshold to educability. The oldest constituent capabilities amounted to some learning and reasoning skills that enabled individuals to learn from their separate experiences so that they could adapt their behaviors to the ever-fluctuating physical conditions they faced. The great strength of such learning systems was that they were grounded in physical reality.

The benefit that educability provided to humans is that the learning and reasoning capabilities that had evolved earlier to cope with the physical world could now be applied to the much richer world of abstract concepts. Abstract systems of beliefs could be developed and communicated within a community. This allowed humans to acquire belief systems that they had not personally developed, and that they had not validated through their own individual experience.

This development, however, came at a price. Humans had acquired the skills to absorb and apply belief systems that pertained to things beyond their personal experience. If survival in the physical world had been the guiding criterion for adopting or adapting behaviors, no criterion that was similarly protective of the individual or species accompanied this new development.

It may be that our facility to create belief systems is so great that no critical judgment can hold it in check. At the time of the Black Death pandemic that devastated the populations of Europe and Asia in the fourteenth century, people sought to understand the cause. In response to a request by King Philip VI of France, a report in 1348 by the Medical Faculty of the University of Paris concludes with the following well-referenced passage:

> We say that the distant and first cause of this pestilence was and is the configuration of the heavens. In 1345, at one hour after noon on 20 March, there was a major conjunction of three planets in Aquarius. This conjunction, along with other earlier conjunctions and eclipses, by causing a deadly corruption of the air around us, signifies mortality and famine— and also other things about which we will not speak here because they are not relevant. Aristotle testifies . . . that mortality of races and the depopulation of kingdoms occur at the conjunction of Saturn and Jupiter, for momentous events then arise, their nature depending on the trigon in which the conjunction occurs. And this is found in ancient philosophers, and Albertus Magnus in his book, Concerning the causes of the properties of the elements (treatise 2, chapter 1) says that the conjunction of Mars and Jupiter causes a great pestilence in the air, especially when they come together in a hot, wet sign, as was the case in 1345. For Jupiter, being wet and hot, draws up evil vapors from the earth and Mars, because it is immoderately hot and dry, then ignites the vapors, and as a result there were lightnings, sparks, noxious vapors and fires throughout the air.[3]

Geoffrey de Meaux, writing in Oxford around 1350, did not go along with all this. He thought the total eclipse of the Moon two days earlier was important, too.

Exploiting our facility with beliefs, arbitrary superstitions can now spread through human communities as easily as Newton's Laws. The mental life of societies can be driven by abstractions piled on top of each other and divorced from the world of the senses. Theories of politics, religion, and identity that cannot be verified or falsified against experience can take hold of entire civilizations, as can superstitions and myths. For millions of years evolution had selected for behaviors that favored survival in the physical world. Through educability, these safe behaviors have now suddenly acquired strong competition from ungrounded ideologies. The evolutionary protections of being grounded in the physical world and self-preserving behaviors are gone.

How to exploit our facility with theories, to our benefit but not to our detriment, is the most consequential challenge we face as humans. The educability framework provides a perspective on it. But there is no simple solution. Saying that we should only follow "evidence-based" or "fact-based" theories goes only so far. After all, in December 2020, in the middle of the Covid-19 pandemic, we had another Jupiter-Saturn conjunction, an even closer one than occurred in March 1345 and was cited by the Medical Faculty of the University of Paris in 1348 as the cause of the Black Death.

The Case of Isaac Newton

The idea that there is a continuous thread of progress that has led us from widespread human sacrifice to the triumph of modern science and technology is not convincing. Humans may have extraordinary facility for creating and transmitting belief systems, but our judgment in evaluating them is weak—too weak, in my opinion, to guarantee progress. It is easy to fall for beliefs that look preposterous to succeeding generations.

The case of Isaac Newton is noteworthy. His ability to create fundamental scientific theories is unsurpassed in human history. Through his theories he was uniquely impactful both in creating physical theories and in defining by example what later generations considered to be valid science. The science he is now so famous for, however, occupied only a small part of his theory-creating efforts. He applied much of his energies to areas such as alchemy and the occult that would now be considered bogus from any academic standpoint. The voluminous manuscripts that he left on these subjects were little wanted, little read, and generally embarrassing to his admirers down the centuries. These writings are "evidence-based" in that they refer to experiments, historical dates, the Bible, and the dimensions of the Egyptian pyramids. They include predictions of the end of the world, 2060 being one date he mentions. They include about a million words on alchemy; he was much concerned with the philosopher's stone for turning base metal into gold.

Later generations concluded that the validity of his science was different from that of his other writings. I suggest that these differences arise from differences in how these theories relate to reality, rather than any differences in how he was processing them in his brain. He did not have a split mind. He was not half mad.

Belief Systems: Good and Bad

This volume suggests that the acquisition and processing of belief systems is appropriate for scientific attention as a technical field of study. But where does that leave the ethical side? Is the evolution of widely held belief systems a mechanistic process

divorced from ethics? Are there any natural guardrails that ensure that the belief systems humans adopt have a tendency to
be good or to be improving?

These questions I will not be able to answer, but history suggests that there are again reasons for concern. Natural ethical
guardrails, if they exist, do not appear to be that sturdy. There
are examples of ethical beliefs once widely and strongly held
that have been largely discarded. One is the prohibition of
usury, the charging of any or excessive interest on loans. There
have been prohibitions on it in Buddhism, Judaism, Christian-
ity, and Islam. Many philosophers have disapproved of it. It was
and still is legally regulated in many jurisdictions. One might
have thought that the breadth of this prohibition and disap-
proval suggests a natural moral law in the background. Yet I
keep getting credit card offers with 30 percent annual interest
rates clearly stated, and the ethics around this does not attract
much attention from present-day leaders.

For an understanding of the benefits and dangers of belief
systems, a technical, ethics-free analysis may have some value.
In this volume I have provided one for the processing of belief
systems. This is different from analyzing the content of belief
systems themselves, which is more traditional. It is not difficult,
for example, to detect when two belief systems are in conflict. For
example, if two nations teach their populations that they are
entitled to the maximum territory that their nation enjoyed in
recorded history, then it seems clear that conflict will follow.

As discussed earlier, the basis of the educability model is the
phenomenon of learning from experience, which is already
present in much simpler animals, such as the sea snail. In those
simpler cases, the learning system will seek to track the external
physical realities that are experienced. Humans, however, with
our symbolic naming capabilities, face more serious challenges.

If one has been taught some abstractions and some arbitrary rules that relate them, then one has internalized beliefs with no guaranteed connection to any external reality. By imagining scenes composed of these abstractions in one's Mind's Eye, one can learn new generalizations about this abstract world to add to what one has been taught. The more one thinks in the Mind's Eye about this imaginary world, the richer the theory one develops, and the further this theory may drift from any external reality or the beliefs of others.

A form of verification is spot-checking for internal consistency. One can apply the rules in different ways to a situation, and if one gets contradictory predictions, then one has a warning sign that the rules are inconsistent. The situations tested may be provided by the external environment or generated internally. Widely circulating belief systems, including conspiracy theories and superstitions, often have substantial internal consistency and are well-inoculated against this sanity check.

Among self-consistent theories are scientific theories, for which no real-world counterexample has been found after decades or centuries of systematic efforts to find one. These offer the pinnacles of the power of educability. But many other rule sets, self-consistent as they may also appear to be, will make predictions that are not verifiable or falsifiable against any outside reality. Such rule sets may relate to almost any field of human interest, real or fictitious.

The basic capability of educability is challenged when differentiating between great scientific theories and baseless myths. In both cases the beliefs may be self-consistent. In both cases it may not be feasible for the individual to make direct checks against reality. For example, for most facts about science, we need to rely on the reports of others. In such instances, the belief choice aspect of educability becomes essential.

Recognizing that we process different belief systems in our brains in similar ways does not mean that they are to be treated as equivalent. There is no contradiction in an individual differentiating among belief systems as being worthy or not of their support.

Science as a Belief System

Fluency with belief systems gave humans the opportunity to develop the advanced technological civilization that we have. While the power of science is for all to see, the reasons it has been so productive in the past are not so self-evident. Even less obvious is what we need to do to keep the benefits of science flowing in the future.

Science is a belief system. A critical component of it is the idea that there are important patterns in the world that are not readily visible but are worth the effort to discover. That there is a moderate number of chemical elements and all materials on Earth are composed of these is not self-evident and had to be demonstrated through ingenuity and labor. Similarly, the bacterial and viral causes of disease are not obvious to the eye. The individuals who made these discoveries believed that useful but well-hidden patterns existed and could be found.

Science as a belief system, even when pursued by imperfect self-interested individuals, has strong self-correcting tendencies. An announcement of any significant result will prompt other scientists to seek to verify it by repeating the experiment or analysis. Also, there is usually wide agreement on the interpretation of an experimental result or an analysis with respect to a relevant question. These two facts in tandem keep the scientific enterprise on track, despite all the mistakes that may occur along the way. They also keep the incidence of fraud to a low level and the influence of any one such fraudulent act usually to a short duration.

We need to take on trust that experiments carried out by others gave the claimed results. We also have to trust that analyses give the results stated without us having to repeat them. To support this trust, there is a complex infrastructure of authoritative textbooks, peer review, research journals, and other publication venues. From the outside, which sources are most trustworthy may be hard to discern, but experts mostly know how much to trust any source in their own field. In the practice of science, where so much knowledge is transferred by instruction, scientists need and do practice sophisticated, but not foolproof, belief choice policies.

Press articles and news releases about science do not go through rigorous review. Nevertheless, they play significant roles in informing the broader public. Such articles and releases rarely mislead scientists, who understand that newsworthy science is often unsettled science. At any time, there is substantial consensus among experts on the settled core of each scientific field. The areas that are not settled and recognized as such are the subjects of most of the ongoing research.

The reasons science works so well are more to do with the world to which it applies than the particular way humans approach it. Science might be compared to a gold mine where each field of science is a vein. Near each vein of well-established science, so much more new science can be unearthed. Venturing beyond these known productive veins gives less predictable results but can lead to new even more productive veins.

No human activity or judgment occurs in isolation. They all occur in the context of some belief system. For any judgment or activity, one ought to declare the context in which it can be justified. For example, the perspective of this volume is the science belief system from a computational perspective.

The Scientific Revolution

While the gift of educability may yet bring humanity to its destruction, it has also led to great triumphs. Educability is a capacity that has taken a long time to have its impact. The possibility of accumulating knowledge discovered by others and creating new knowledge from it may have existed for hundreds of thousands of years. Eventually this gift spectacularly caught fire as the Scientific Revolution, which unfolded in the sixteenth and seventeenth centuries. Its principal protagonists lived in different corners of Europe, employed variously by universities, rulers, and religious entities, or living on personal wealth. They had Latin as a common language. They published their work in printed books, a technology invented not much earlier. They read one another's work. Clubs and meeting places arose to bring together local groups of scientific researchers in Italy, Spain, England, Germany, and France. By the late seventeenth century, these had evolved into academies, including the Royal Society in London, the Académie des Sciences in Paris, and the Leopoldina in Germany.

Why this unique event, the Scientific Revolution, took place exactly when and where it did, some three hundred thousand years after the emergence of our species, is open to debate. The precipitating event was not a genetic mutation in fifteenth-century Europe. The event often cited as the one that unleashed this development was the invention of the Gutenberg printing press. Around 1438 Johannes Gutenberg, living in Strasbourg, invented a printing press that made the production of books much less laborious and expensive than before. His was a movable-type press, in which the individual letters were made of metal and manufactured separately. One could assemble these letters rapidly to make a page of print and later reuse them to make another. In this invention, which incorporated many previously existing ideas, Gutenberg made several innovations that revolutionized the production of books. His ideas spread rapidly. They led to more affordable books, a substantial book trade, and an increasingly literate population. Since the dissemination of theories is at the heart of educability, it is no surprise that a technology that so revolutionized this dissemination was a great spur to human civilization.

Perhaps it was inevitable that humanity would eventually stumble on the Scientific Revolution, even though it may have had to wait for a fortuitous combination of circumstances. The movable-type printing press had been invented in China in the eleventh century and used in Korea in the thirteenth century. In East Asia, however, movable-type printing did not displace woodcut printing or writing by scribes until centuries later. There are various theories for why movable type did not have such dramatic effects there. One theory is that in European languages where the alphabet size was small, the cost reduction offered by movable-type printing was much more dramatic than in East Asia, where alphabets were much larger.

In the centuries since, scientific research has become more institutionalized and professionalized. Currently doctoral students in any number of disciplines train to absorb knowledge in highly specialized areas so that they can further advance their fields. Humanity is finally exploiting the gift of educability in a systematic way and on an industrial scale. At the same time, it is enjoying all the benefits the new scientific knowledge provides.

It is quite possible that further improvements can be made in the scientific research process itself.[4] A scientist needs access to previous knowledge, convenient ways of isolating those pieces that are relevant to the research question at hand, and new ideas. The sharing of information on the web and the use of search engines have already had important effects. Scientists can now more rapidly follow what is happening in their field, which will have orders of magnitude more participants than Kepler or Newton had to follow. Digital technology may well be launching a phase of scientific progress that is even more intense than before. This is not just because of all the opportunities computers offer for simulating scientific theories and detecting patterns in data, but more simply because digital technology offers another revolution in the dissemination of knowledge.

Equality

That "all men are created equal" was "sacred and undeniable" to Thomas Jefferson in his draft of the United States Declaration of Independence. With later editing, it became "self-evident" in the final document. While historians still debate Jefferson's own intent, the continuing impact of his phrase prompts the question of how to interpret the words now. Can one justify Jefferson's final wording as a statement of fact? Several religions support the concept of equality, and hence Jefferson's choice of "sacred" would be more understandable. At no point in history, however, has equality been self-evident from looking around. Different social classes in the same region and the same classes in different regions have had different enough lives to make such a proposition counterfactual on the surface.

I suggest that the educability hypothesis fills a gap here in providing an angle from which to view our equality. Educability implies that humans, whatever our genetic differences at birth, have a unique capability to transcend these differences through the knowledge, skills, and culture we acquire after birth. We are born equal because any differences we have are subject to enormous subsequent changes through individual life experience, education, and effort. This capacity for change, growth, and improvement is the great equalizer. It is possible for billions of people to continuously diverge in skills, beliefs, and knowledge, all becoming self-evidently different from each other. This characteristic of our humanity, which accounts for our civilization, also makes us equal.

Perhaps the most serious challenge to the notion of human equality that came from modern science was the eugenics movement. The term comes from the Greek word for "well-born." The adherents believed that inequality at birth was fundamental. The movement flourished from the 1880s to the 1930s in Europe and North America, driven by the eugenicists' fear that if people with so-called "superior" genes reproduced at a lower rate than those with "inferior" genes, then humanity's genetic stock would decline. The eugenicists proposed to take measures to discourage reproduction of what they considered the "inferior" genes. The criteria they considered as valid to discriminate between superior and inferior included measures such as IQ scores as well as membership of national and racial groups.

The reason for eugenics eventually falling into disrepute was neither the ethical issue it obviously raised nor scientific questioning, but rather the wide-ranging use of its tenets by the Nazi regime in Germany. Subsequently, in 1948 the United Nations General Assembly adopted the Genocide Convention, which included in its definition of genocide the imposition of any measures intended to prevent births within a "national, ethnical, racial or religious" group.

The history of eugenics deserves study as an example where the self-correcting tendency of science was not in evidence for a long time. Several of the creators and primary movers of eugenics were among the most prominent scientists of their time. Some of them laid the foundations of modern statistics. Users

of statistics will recognize their names and the statistical techniques they contributed: Francis Galton (regression to the
mean, standard deviation), Karl Pearson (chi-squared test, principal component analysis), and Ronald Fisher (tests of significance, maximum likelihood testing).

It is now widely thought that these statisticians were mistaken
in their belief in the primacy of human inequality at birth. How
could such a group have been so wrong? Their published papers
were chock-full of data, and they were applying their scientific
expertise—statistics—for analyzing the data. Their statistical
methodologies continue to be the basis for understanding data
in the empirical sciences to this day. My suggestion is that the
data available to them did not provide much information about
the power of educability. They had little systematic data on
individuals of different classes, cultures, ethnicities, and gender being subject to the same educational opportunities over
an extended period. Human educability is a surprising and
amazing phenomenon that these statisticians—and their
contemporaries—failed to detect. Some would explain the eugenics movement by saying simply that the participants were
biased by the beliefs of their times. The power of beliefs is, of
course, a central theme of this book—but beliefs come from somewhere, and one should try to understand their origin.

For many measurable traits of plants and animals, scientists
have sought to distinguish the effects of nature and nurture by
assigning percentages to each of these sources using statistical
analysis. Such analysis by itself provides no understanding of
the mechanisms by which nature and nurture influence the
trait. For cognitive traits, such analysis is particularly problem-
atic if we accept that education plays a role in the development
of the trait. Educability allows the influence of the environment
to be quite enormous. The essence of educability is the unique
and extreme power of this influence.

The answer to the eugenicists' fear is that the capacity to change through experience and education is at the very center of the architecture of the human mind. Fortunately for humanity, the main social development of the last century has been the worldwide expansion of education. We are still in the middle of this expansion. Improvements in education and in the numbers receiving it have overwhelming potential. Pursuing this potential is the most rewarding focus for anyone aiming to improve a society.

Worldviews

Belief systems that are broad enough to suggest positions on diverse issues have been called worldviews. Religions are examples. Political and economic systems are others. Throughout human history there have always been widely held and wildly different belief systems about race, class, and gender, about who is the enemy, and about who is fully human.

Worldviews continue their struggle for acceptance every day. In each epoch various worldviews have been particularly influential. It is customary to be smug about one's own worldview and dismissive of those of others, especially those of earlier times. Educability offers, among other things, an onlooker's vantage point on this struggle.

Currently, a worldview with much influence is science. I think this is good. Nations that include most of the human population are teaching science to their young and putting resources into exploiting it for the benefit of their citizens. The proven success of this enterprise is one thing that we can be sure about and would do well to further. A scientific consensus on the nature of the Civilization Enabler may be impactful. Some convergence on what defines us may promote commonality among the ruling worldviews.

We therefore have opportunities.

What about the threats? What guarantees do we have that the ruling worldviews will not follow trajectories that most would currently regard as undesirable, such as returning to widespread human sacrifice? The answer seems to be none. Humans are just too facile with absorbing and applying arbitrary belief systems, and we seem to have much weaker countervailing abilities to evaluate the consequences or validity of our beliefs. This volume shows that one can discuss aspects of belief system acquisition and processing with some precision. Much remains to be discovered. What makes an individual commit to a belief system? What makes an individual maintain their commitment or give it up when it is challenged?

There is little evidence that our civilization is securely on an upward slope. Wars and oppressive political regimes persist and continue to attract supporters. I see no guarantee that belief systems that are worse, or much worse, than those that currently dominate around the world will not displace them. The only actionable defense I see is to seek a better understanding of how we process beliefs. Understanding our critical capacity could help guard against its worst dangers.

Educability is an information processing capability that humans have. It has enabled us to stand on each other's shoulders and build the edifice of beliefs that is our current civilization. Our power to generate and adopt new beliefs has few limits. The beliefs we adopt govern how we act and have consequences without end. One would hope that if educability is recognized to be the defining human capability, then the search for a deeper understanding of it would become a unifying quest. Surely, we can direct our power of being educable at ourselves, to better understand the nature of this power and set a steadier course.

ACKNOWLEDGMENTS

This volume has greatly benefited from input from several individuals who have gone well beyond the call of duty. It is a pleasure to acknowledge helpful comments on early versions of the manuscript from Helen Chambers and Juliet Harman, valuable advice from Paul Harris, helpful conversations with Nick Patterson and David Reich, and useful feedback from three anonymous referees. I would also like to thank my editor Ingrid Gnerlich at Princeton University Press for investing so much of her time in this project. My wife, Gayle Valiant, contributed to more aspects of this enterprise than I can describe, and I thank her.

SUMMARY OF TERMINOLOGY

Some terms in the text are used in a technical sense. This sense is defined in the text in intuitive terms for the general reader. The pages where these descriptions are to be found are italicized in the index. More precise definitions of the basic concepts can be found in the notes or references therein. This summary is a guide to the main technical terms.

Educability is the primary technical term. It describes a capability that is the subject of this book. A high-level definition is in terms of the three pillars: (a) learning from experience, (b) being teachable by instruction, and (c) combining and applying what has been obtained in both modes. The detailed technical definition is the subject of chapters 6 and 8. Educability is equivalent to an *Educable Learning System,* to the extent that the latter is a description of a system that has the behavioral capability of the former.

An Educable Learning System (ELS) is an Integrative Learning System with the two additions that it can *acquire rules by instruction* and can fully handle *symbolic naming.* A *rule acquired by instruction* is like a cookery recipe described explicitly or an executable program. (It is not learned by example.) Symbolic naming (which does not have a mathematical definition) is a general scheme to describe things by means, such as by the words of a human language, which are arbitrary and not related to the things described.

An Integrative Learning System (ILS) is a system that can learn from examples by generalization and can *chain* together these generalizations to classify examples it has not seen before. An integral part of it is a *Mind's Eye*, which is a device to which the examples are presented. In the Mind's Eye each example is presented as a *scene*, and the different components of the scene and the relationships among the components can be examined. Each component of an example will be represented by a *token*, and each relationship by an *attribute*. An attribute can relate to one or more tokens. The memory contains a set of *rules* that contain the stored knowledge and are used to update individual scenes. The rules may have been *learned* by generalization from examples (in the case of an ILS or ELS) or *acquired by instruction* (in the case of an ELS).

A *generalization* is a rule, like a cookery recipe, but one learned from examples by a *learning algorithm*. Such a learning algorithm processes a number of examples, each labeled according to some criterion, such as whether the example picture represents a certain animal. The algorithm outputs a *classifier* or rule that can predict the criterion for new examples. The *features* of a classifier are the inputs in terms of which examples are specified, such as the pixels in the case that the examples are images. The *target* is the criterion according to which the classification is to be made. When the learning algorithm is given some examples from which to learn, each such example comes with a *label* that gives the value of the target for that example. When a scene of the Mind's Eye is an example, the features and targets are expressed in terms of tokens and attributes.

Probably Approximately Correct (PAC) learning is a quantitative condition on learning algorithms that needs to be satisfied for the algorithms to be regarded as capable of learning. The condition is concerned with the accuracy of the predictions of the classifier on new examples and on the feasibility of the computation performed by the learning algorithm.

NOTES

Chapter 1

1. Alyn Cairns, private communication, 9/30/2019. Also https://www.bbc.com /news/uk-northern-ireland-47186124, accessed 5/11/2023.

2. Reviews of "Aesop Fable" phenomena: S. Ghirlanda and J. Lind, "'Aesop's Fable' Experiments Demonstrate Trial-and-Error Learning in Birds, but No Causal Understanding," *Animal Behavior* 123 (2016): 239–46; L. Hennefield et al., "Meta-analytic Techniques Reveal that Corvid Causal Reasoning in the Aesop's Fable Paradigm Is Driven by Trial-and-Error Learning," *Animal Cognition* 21 (2018): 735–47.

3. S. M. Reader and K. N. Laland, eds., *Animal Innovation* (Oxford: Oxford University Press, 2003); A. N. P. Stevens and J. R. Stevens, "Animal Cognition," *Nature Education Knowledge* 3, no. 11 (2012): 1; A. B. Kaufman and J. C. Kaufman, *Animal Creativity and Innovation* (Amsterdam: Elsevier, 2015).

4. R. Nielsen et al., "Tracing the Peopling of the World through Genomics," *Nature* 302 (2017): 541; D. Reich, *Who We Are and How We Got Here: Ancient DNA and the New Science of the Human Past* (New York: Pantheon, 2018).

5. S. Mallick et al., "The Simons Genome Diversity Project: 300 Genomes from 142 Diverse Populations," *Nature* 538 (2016): 201.

6. The following study gives evidence that major changes in the vocal and facial anatomy in human ancestors were realized by epigenetic as opposed to genetic changes: D. Gokhman et al., "Differential DNA Methylation of Vocal and Facial Anatomy Genes in Modern Humans," *Nature Communications* 11 (2020): 1189. The epigenome consists of mechanisms outside of the DNA that also regulate the expression of proteins. The epigenome is both heritable and subject to environmental influences.

7. I. Gronau et al., "Bayesian Inference of Ancient Human Demography from Individual Genome Sequences," *Nature Genetics* 43 (2011): 1031–34; K. R. Veeramah et al., "An Early Divergence of KhoeSan Ancestors from Those of Other Modern Humans Is Supported by an ABC-Based Analysis of Autosomal Resequencing Data," *Molecular Biology and Evolution* 29 (2012): 617; C. M. Schlebusch et al., "Genomic

Variation in Seven Khoe-San Groups Reveals Adaptation and Complex African History," *Science* 338 (2012): 374–79; C. M. Schlebusch et al., "Khoe-San Genomes Reveal Unique Variation and Confirm the Deepest Population Divergence in Homo sapiens," *Molecular Biology and Evolution* 37, no.10 (2020): 2944–54.

8. M. F. Hammer et al., "Genetic Evidence for Archaic Admixture in Africa," *PNAS* 108, no. 37 (2011): 15123–28; J. K. Pickrell, "Ancient West Eurasian Ancestry in Southern and Eastern Africa," *PNAS* 111, no. 7 (2014): 2632–37.

9. Donald Knuth has suggested that the identifying characteristic of computer science as compared with mathematics is the notion of state: D. E. Knuth, "*Algorithms in Modern Mathematics and Computer Science*," *Lecture Notes in Computer Science*, vol. 122 (Berlin: Springer, 1979), 82–99.

10. The talk was based on the coauthored paper by H. A. Simon and A. Newell, "Heuristic Problem Solving: The Next Advance in Operations Research," *Operations Research* 6, no.1 (1958): 1–10.

Chapter 2

1. C. Darwin, *The Descent of Man, and Selection in Relation to Sex* (London: Murray, 1871).

2. D. Campbell, "Variation and Selective Retention in Socio-cultural Evolution," in *Social Change in Developing Areas: A Reinterpretation of Evolutionary Theory*, ed. H. Barringer, G. Blanksten, and R. Mack (Cambridge, MA: Schenkman, 1965), 19–49; M. Feldman and L. Cavalli-Sforna, "Cultural and Biological Evolutionary Processes, Selection for a Trait under Complex Transmission," *Theoretical Population Biology* 9, no. 2 (1976): 238–59; C. Lumsden and E. Wilson, *Genes, Mind and Culture: The Coevolutionary Process* (Cambridge, MA: Harvard University Press, 1981); A. Whiten, "The Burgeoning Reach of Animal Culture," *Science* 372, no. 6537 (2021).

3. L. G. Valiant, *Circuits of the Mind* (Oxford: Oxford University Press, 1994); V. Feldman and L. G. Valiant, "Experience-Induced Neural Circuits That Achieve High Capacity," *Neural Computation* 21, no. 10 (2009): 2715–54.

4. J. B. Smaers et al., "Exceptional Evolutionary Expansion of Prefrontal Cortex in Great Apes and Humans," *Current Biology* 27 (2017): 714–20; C. J. Donahue et al., "Quantitative Assessment of Prefrontal Cortex in Humans Relative to Nonhuman Primates," *PNAS* 115, no. 22 (2018): E5183–92.

5. J. K. Rilling et al., "The Evolution of the Arcuate Fasciculus Revealed with Comparative DTI," *Nature Neuroscience* 11 (2008): 426–28.

6. T. M. Preuss, "The Human Brain: Rewired and Running Hot," *Annals of the New York Academy of Science* 1225 (2011): E182–91.

7. M. S. Ponce de Leon et al., "The Primitive Brain of Early *Homo*," *Science* 372 (2021): 165–71.

8. S. Neubauer, J. J. Hublin, and P. Gunz, "The Evolution of Modern Human Brain Shape," *Science Advances* 4, no. 1 (2018): 24. Among possible evolutionary mechanisms for this are yet undetected genetic mutations, polygenic adaptations, or epigenetic changes.

9. MOCA topics, Center for Academic Research and Training in Anthropogeny, https://carta.anthropogeny.org/moca/topics, accessed 5/11/2023.

10. J. V. Lawick-Goodall, "Tool Using in Primates and Other Vertebrates," in *Advances in the Study of Behaviour*, vol. 3, ed. D. S. Lehrman, R. A. Hinde, and E. Shaw (Cambridge, MA: Academic Press, 1970); C. Rutz et al., "Discovery of Species-Wide Tool Use in the Hawaiian Crow," *Nature* 537 (2016): 403–7.

11. C. Rutz et al., "Tool Bending in New Caledonian Crows," *Royal Society Open Science* 3 (2016).

12. C. Rutz and J. H. St. Clair, "The Evolutionary Origins and Ecological Context of Tool Use in New Caledonian Crows," *Behavioural Processes* 89, no. 2 (2012): 153–65.

13. A. P. M. Bastos et al., "Self-Care Tooling Innovation in a Disabled Kea (*Nestor notabilis*)," *Scientific Reports* 11 (2021): 18035.

14. D. T. Ksepka et al., "Tempo and Pattern of Avian Brain Size Evolution," *Current Biology* 30 (2020): 1–11.

15. K. Laland and A. Seed, "Understanding Human Cognitive Uniqueness," *Annual Review of Psychology* 72 (2012): 689–716.

16. P. J. Richerson and R. Boyd, *Not by Genes Alone: How Culture Transformed Human Evolution* (Chicago: University of Chicago Press, 2005); K. N. Laland, *Darwin's Unfinished Symphony: How Culture Made the Human Mind* (Princeton, NJ: Princeton University Press, 2017).

17. J. Henrich, *The Secret of Our Success* (Princeton, NJ: Princeton University Press, 2018).

18. M. Tomasello, *Becoming Human: A Theory of Ontology* (Cambridge, MA: Belknap Press, 2019).

Chapter 3

1. G. K. Chesterton, as quoted in "Sayings of the Week," *Observer* (London), July 6, 1924.

2. Tom Nichols, radio interview, *On Point*, WBUR, March 17, 2020.

3. Aristotle, *Politics*, book 8, part 1.

4. U. Geisser, "Intelligence: Knowns and Unknowns," *American Psychologist* 51, no. 2 (1994): 77–101. The reference to the two dozen opinions is R. J. Steinberg and D. K. Detterman, *What Is Intelligence?* (Westport, CT: Praeger, 1986).

5. C. Spearman, "General Intelligence Objectively Determined and Measured," *American Journal of Psychology* 15, no. 2 (1904): 201–93.

Chapter 4

1. Various versions attributed to Isidore of Seville, Desidarius Erasmus, and Mahatma Gandhi.

2. D. L. Alkon, "Learning in a Marine Snail," *Scientific American* 249, no. 1 (1983): 70–83; T. Crow, "Pavlovian Conditioning of Hermissenda: Current Cellular, Molecular, and Circuit Perspectives," *Learning & Memory* 11 (2004): 229–37.

3. R. J. Herrnstein et al., "Natural Concepts in Pigeons," *Journal of Experimental Psychology: Animal Behavior Processes* 2, no. 4 (1976): 285–302.

4. M. J. Morgan et al., "Pigeons Learn the Concept of an 'A,'" *Perception* 5, no. 1 (1976): 57–65.

5. S. Gelman, "Learning from Others: Children's Construction of Concepts," *Annual Review of Psychology* 60 (2013): 115–40.

6. The concept of PAC learning was introduced in L. G. Valiant, "A Theory of the Learnable," *Communications of the ACM* 27, no. 11 (1984): 1134–42. That paper used the word "learnable" to describe what could be learned. The PAC name was later suggested in D. Angluin and P. Laird, "Learning from Noisy Examples," *Machine Learning* 2 (1988): 343–70. Note that the term "PAC learning" is sometimes used in a purely statistical sense (e.g., V. Vapnik, *Statistical Learning Theory* [New York: Wiley, 1998]), where arbitrarily large and infeasible amounts of computational effort are permitted in the learning process. This purely statistical sense would be too permissive for the current work, which studies phenomena that are computationally feasible to realize in this world.

7. L. Standing, "Learning 10,000 Pictures," *Quarterly Journal of Experimental Psychology* 25, no. 2 (1973): 207–22.

8. World Memory Championships, https://en.wikipedia.org/wiki/World _Memory_Championships, accessed 7/21/2023.

9. S. B. Vander Wall, *Food Hoarding in Animals* (Chicago: University of Chicago Press, 1990).

10. J. Fagot and R. G. Cook, "Evidence for Large Long-Term Memory Capacities in Baboons and Pigeons and Its Implications for Learning and the Evolution of Cognition," *PNAS* 103, no. 46 (2006): 17564–67.

11. A. Saito et al., "Domestic Cats (*Felis catus*) Discriminate Their Names from Other Words," *Scientific Reports* 9 (2019): 5394.

12. S. L. King and V. M. Janik, "Bottlenose Dolphins Can Use Learned Vocal Labels to Address Each Other," *PNAS* 110, no. 32 (2013): 13216–21.

13. E. Hess, *Nim Chimpsky: The Chimp Who Would Be Human* (New York: Bantam Books, 2008).

14. F. Patterson and E. Linden, *The Education of Koko* (New York: Holt, 1983).

15. J. W. Pilley and A. K. Reid, "Border Collie Comprehends Object Names as Verbal Referents," *Behavioural Processes* 86, no. 2 (2011): 184–95.

16. T. M. Caro and M. D. Hauser, "Is There Teaching in Nonhuman Animals?" *Quarterly Review of Biology* 67, no. 2 (1992): 151–73.

17. S. Musgrave et al., "Tool Transfers Are a Form of Teaching among Chimpanzees," *Scientific Reports* 6 (2016): 34783.

18. N. R. Franks, "Teaching in Tandem-Running Ants," *Nature* 439 (2006): 153.

19. A. Thornton and K. McAuliffe, "Teaching in Wild Meerkats," *Science* 313 (2006): 227–29.

20. L. A. Ogden, "Do Golden Lion Tamarins Teach Their Young?" *BioScience* 61, no. 11 (2011): 932.

21. P. Reps and N. Senzaki, *Zen Flesh, Zen Bones: A Collection of Zen and Pre-Zen Writings*, Tuttle Publishing, 1998.

22. E. Herrmann et al., "Humans Have Evolved Specialized Skills of Social Cognition: The Cultural Intelligence Hypothesis," *Science* 317 (2007).

23. M. Tomasello, S. Savage-Rumbaugh, and A. C. Kruger, "Imitative Learning of Actions on Objects by Children, Chimpanzees and Enculturated Chimpanzees," *Child Development* 64, no. 6 (1993): 1688–1705.

24. E. Bonawitz et al., "The Double-Edged Sword of Pedagogy: Instruction Limits Spontaneous Exploration and Discovery," *Cognition* 120, no. 3 (2011): 322–30.

25. R. A. Rescorla, *Pavlovian Second-Order Conditioning: Studies in Associative Learning*, Psychology Revivals (Hove, UK: Psychology Press, 2015); J. C. Gewirtz and M. Davis, "Using Pavlovian Higher-Order Conditioning Paradigms to Investigate the Neural Substrates of Emotional Learning and Memory," *Learning and Memory* 7 (2000): 257–66.

26. Another example of chaining: J. H. Wimpenny et al., "Cognitive Processes Associated with Sequential Tool Use in New Caledonian Crows," *PLoS One* 4, no. 8 (2009): e6471.

27. G. A. Miller, "The Magical Number Seven, Plus or Minus Two: Some Limits on Our Capacity for Processing Information," *Psychological Review* 63, no. 2 (1956): 81–97.

28. L. G. Valiant, *Probably Approximately Correct* (New York: Basic Books, 2013), 125–29.

29. W. Kohler, *The Mentality of Apes*, trans. Ella Winter (London: Kegan Paul, Trench, Trubner, 1925).

Chapter 5

1. "Turing Machine," https://en.wikipedia.org/wiki/Turing_machine, accessed 5/11/2023. The original paper is A. M. Turing, "On Computable Numbers, with an Application to the Entscheidungsproblem," *Proceedings of the London Mathematical Society* 42 (1936–1937): 230–65.

2. In L. G. Valiant, "A Theory of the Learnable," *Communications of the ACM* 27, no. 11 (1984): 1134–42, it was shown, for example, that the class of 3CNF Boolean

formulae was PAC learnable. For this class, determining whether there exists any input for which the value of the formula is 1 is NP-complete. This is no impediment to learning, which requires good behavior only on typical inputs and not on exponentially rare ones. See also M. J. Kearns and U. V. Vazirani, *An Introduction to Computational Learning Theory* (Cambridge, MA: MIT Press, 1994); M. Mohri, A. Rostamizadeh, and A. Talwalkar, *Foundations of Machine Learning*, 2d ed. (Cambridge, MA: MIT Press, 2018).

3. Empirical evidence for deep learning having PAC scaling: J. Hestness et al., "Deep Learning Scaling Is Predictable, Empirically," https://arxiv.org/abs/1712 .00409v1. Errors decreasing as fixed exponents of the effort are sometimes called "power laws."

4. E. Strubell, A. Ganesh, and A. McCallum, "Energy and Policy Considerations for Deep Learning in NLP," *arXiv*, 1906.02243 (2019); D. Patterson et al., "Carbon Emissions and Large Neural Network Training," *arXiv*, 2104.10350 (2021).

5. D. Kahneman, *Thinking, Fast and Slow* (New York: Farrar, Straus and Giroux, 2011).

6. R. E. Rubin and J. Weisberg, *In an Uncertain World: Tough Choices from Wall Street to Washington* (New York: Random House, 2004).

7. Here I extend the notion from Valiant, *Probably Approximately Correct*, 28–31, by adding the second, Type B, form of robustness.

8. D. E. Knuth, *The Art of Computer Programming*, vol. 3: *Sorting and Searching* (Boston: Addison-Wesley, 1998).

9. D. Haussler et al., "Equivalence of Models for Polynomial Learnability," *Information and Computation* 95, no. 2 (1991): 129–61.

Chapter 6

1. The following paper gives full technical details for many of the points made in this chapter and the next: L. G. Valiant, "Robust Logics," *Artificial Intelligence Journal* 117 (2000): 231–53. The term "relation" as used there is replaced by "attribute" here. Integrative learning as described in chapter 6 is a way of understanding the functionality that a Robust Logic system achieves. In turn, educability, as described in chapter 8, is an Integrative Learning System that has been enhanced to be able to chain rules acquired by instruction, in addition to learned rules. (As far as terminology, the word "robust" here refers to the fact that this system tolerates errors. The use of this word earlier in "robust models of computation" is different and refers to invariance under change of model.)

2. To represent the information as assertions equivalent to the diagrams but without diagrams, we would give names, such as *a, b, c, d*, to the different tokens. For example, for figure 3, the equivalent assertions could be toy(a), duck(a), big(b),

tiger(b), and on(a, b), and for figure 4, seed(a), water(b), glass(c), water_level(d), floating_on(a, b), in(b, c) and water_level_of(d, c).

3. The distinction between reflex responses and those that require more thought has a long history in psychology: William James, *The Principles of Psychology* (New York: Holt, 1890); S. A. Sloman, "The Empirical Case for Two Systems of Reasoning," *Psychological Bulletin*, 119 (1996): 3–22; D. Kahneman, *Thinking, Fast and Slow* (New York: Farrar, Straus and Giroux, 2011.) The same distinction was made on computational grounds in L. G. Valiant, *Circuits of the Mind* (Oxford: Oxford University Press, 1994): 57–72.

4. The complete formalism would need to again distinguish the different tokens by name and token variable name. Thus we would need to say ForAllTokens x [Condition1 OR Condition2 $\cong$ rises(x)], where x is a variable that can range over all the tokens, which, as we have said, are $a, b, c, d, \ldots$. The meaning of this rule is that for each token, the attribute rises will be predicted to hold for that token if at least one of Condition1 or Condition2 holds in the scene for that same token. Necessarily, Condition1 and Condition2 will involve some attributes on token x, since otherwise no prediction can be reasonably made specific to that token.

5. The fuller specification of Condition1 would be ThereExistsToken y ThereExistsToken z {water_level(x) & water(z) & glass(y) & in(z,y) & stone_added(y) & water_level_of(x,y)}. To recap, the example rule makes an assertion about every token x in the scene. For Condition 1 to hold, it is necessary that water-level(x) holds. If the only token in the scene for which water_level holds is d then Condition 1 can be true only for $x = d$, if at all. If there can be found further tokens y and z that, in combination with the choice $x = d$, satisfy the remainder of Condition 1, then it will be valid to conclude that rises(d) holds, since the rule is deemed correct in the Probably Approximately Correct sense. In this example such tokens can be found, namely, $z = b$ and $y = c$.

A noteworthy aspect of the formulation is that if a classifier has multiple arguments (here an OR has the two arguments Condition1 and Condition2) and if the arguments have quantifiers (here Condition1 has ThereExistsToken y ThereExistsToken z), then the quantifiers of the different arguments are independent of each other. Such arguments are known as Independently Quantified Expressions, or IQEs.

6. R. Reiter and G. Criscuolo, "On Interacting Defaults," *International Joint Conference on Artificial Intelligence* (1981): 270–76.

7. See M. J. Kearns and U. V. Vazirani, *An Introduction to Computational Learning Theory* (Cambridge, MA: MIT Press, 1994).

8. Valiant, *Probably Approximately Correct*, 87–112.

9. Valiant, "Robust Logics," discusses how combinations may be formed.

10. Note that we can express such combinations using the terminology of scenes:

$$\text{ThereExistsToken } y \text{ Leg\_Room}(x, y)\text{Car}(x)\text{Much}(y).$$

This refers to a token x and will be true if and only if the attribute Car holds for x, and there is a second token y in the Mind's Eye for which the two attributes Leg_Room(x, y) and Much(y) both hold. Such an expression asserts that the token x represents a car that has much legroom.

Chapter 7

1. The fuller specification of Robust Logic allows for a second way in which a variable value can be unspecified. This second kind is denoted by *. Here scenes from the data source D come with possible feature values, 0, 1, *. The distribution D captures the fact that the feature values are unspecified in natural cases with certain probabilities, and the * value in an example then provides some information about that example. In contrast, the obscured $ sign has no relation to D and can be an arbitrary choice.

Chapter 8

1. How they came together in this author's mind is easier to identify. It started with PAC learning in 1982–83 and Robust Logic around 1997. The realization that augmenting Robust Logic with teachable rules is transformational in the cognitive capability it provides, is the genesis and novel contribution of this book.

2. C. Tennie, J. Call and M. Tomasello, "Ratcheting Up the Ratchet: On the Evolution of Cumulative Culture," *Philosophical Transactions of the Royal Society B* 364 (2009): 2405–15; S. Yamamoto, T. Humle, and M. Tanaka, "Basis for Cumulative Cultural Evolution in Chimpanzees: Social Learning of a More Efficient Tool-Use Technique," *PLOS One* 8, no. 1 (2013): e55768; A. Whiten, "Social Learning and Culture in Child and Chimpanzee," *Annual Review of Psychology* 68 (2017): 129–54; M. Tomasello, *Becoming Human* (Cambridge, MA: Harvard University Press, 2019).

3. D. E. Knuth, *The Art of Computer Programming*, vol. 2: *Seminumerical Algorithms* (Reading, MA: Addison-Wesley, 1997).

4. A classical choice for this pseudorandom hash function is $y = (ax + b) \bmod c$. Here x is the encoding of the name and y is the memory address to which it maps, both as before. We now also have the three numbers a, b, c, which stay fixed for all the different xs. Here "mod c" means that we divide the value of $ax + b$ by c and take the remainder, which will then be the value of y. This guarantees that y will be a number in the range 0 and $c - 1$. In the instance in question, we would take $c = 3,001$. Also, we choose random values between 1 and 3,000 for a and b, which then stay fixed for all the xs. This randomization will reduce the chances of the worst collisions for any one sequence of xs.

5. Valiant, *Circuits of the Mind*, 72.

6. L. G. Valiant, "Toward Identifying the Systems-Level Primitives of Cortex by In-Circuit Testing," *Frontiers in Neural Circuits*, 12 (November 2018): 104.

7. D. O. Hebb, *The Organization of Behaviour* (New York: Wiley, 1949).

8. Valiant, *Circuits of the Mind*, 75.

9. The contents of working memory, which we associate here with the Mind's Eye, have been formalized in other ways, for example in the context of consciousness (e.g., L. Blum and M. Blum, "A Theory of Consciousness from a Theoretical Computer Science Perspective: Insights from the Conscious Turing Machine," *PNAS* 119 [2022]: 21) or of cognition. Such works discuss what we call here management rules.

10. J. Legge, *Confucian Analects*, xii.7, 254 (Hong Kong, ca. 1895).

11. F. Clément, M. Koenig, and P. L. Harris, "The Ontogenesis of Trust," *Mind & Language* 19, no. 4 (2004): 360–79; M. Koenig, F. Clément, and P. L. Harris, "Trust in Testimony: Children's Use of True and False Statements," *Psychological Science* 15, no. 10 (2004): 694–98; P. L. Harris and K. H. Corriveau, "Young Children's Selective Trust in Informants," *Philosophical Transactions of the Royal Society* B 366 (2011): 1179–87, 2011; I. Schoon and H. Cheng, "Determinants of Political Trust: A Lifetime Learning Model," *Developmental Psychology* 47, no. 3 (2011): 619–31; P. L. Harris, *How Children Learn from Others* (Cambridge, MA: Harvard University Press, 2012).

12. C. Canteloup, W. Hoppitt, and E. van de Waal, "Wild Primates Copy Higher-Ranked Individuals in a Social Transmission Experiment," *Nature Communications* 11 (2020): 459.

13. P. N. Johnson-Laird, "Mental Models and Human Reasoning," *PNAS* 107, no. 43 (2010):18243–50.

14. A. N. Meltzoff, A. Waismeyer, and A. Gopnik, "Learning about Causes from People: Observational Causal Learning in 24-Month-Old Infants," *Developmental Psychology* 48, no. 5 (2012):1215–28; A. Gopnik and H. M. Wellman, "Reconstructing Constructivism: Causal Models, Bayesian Learning Mechanisms, and the Theory," *Psychological Bulletin* 138, no. 6 (2012), 1085–1108.

15. S. Gelman, "Learning from Others: Children's Construction of Concepts," *Annual Review of Psychology* 60 (2013): 115–40.

Chapter 9

1. P. Carruthers, "Evolution of Working Memory," *PNAS* 110 (2013): 10371–78.

2. J. Jordania, *Who Asked the First Question? The Origins of Human Choral Singing, Intelligence, Language and Speech* (Tbilisi, Georgia: Logos Publishing, 2006).

3. A. Luria, *Cognitive Development: Its Cultural and Social Foundations* (Cambridge: Cambridge University Press, 1976).

4. S. Scribner and M. Cole, *The Psychology of Literacy* (Cambridge, MA: Harvard University Press, 1981).

5. M. G. Dias and P. L. Harris, "The Effect of Make-Believe Play on Deductive Reasoning," *British Journal of Developmental Psychology* 6, no. 3 (1988): 207–21.

6. S. Dehaene, *The Number Sense: How the Mind Creates Mathematics* (New York: Oxford University Press, 1997); F. Xu and E. S. Spelke, "Large Number Discrimination in 6-Month-Old Infants," *Cognition* 74, no. 1 (2000), B1–B11; H. Barth, N. Kanwisher, and E. Spelke, "The Construction of Large Number Representations in Adults," *Cognition* 86, no. 3 (2003): 201–21; S. Carey and D. Barner, "Ontogenetic Origins of Human Integer Representations," *Trends in Cognitive Sciences* 23 (2019): 823–35.

7. P. Diaconis and F. Mosteller, "Methods for Studying Coincidences," *Journal of the American Statistical Association* 84, no. 408 (1989): 853.

8. D. B. Rubin, "Estimating Causal Effects of Treatments in Randomized and Nonrandomized Studies," *Journal of Educational Psychology* 66, no. 5 (1974): 688–701; J. Pearl, *Causality: Models, Reasoning, and Inference* (New York: Cambridge University Press, 2000).

9. H. E. Gardner, *Frames of Mind: The Theory of Multiple Intelligences* (New York: Basic Books, 1983).

10. A statement to this effect is attributed, for example, to Howard Aiken in a talk in 1954, in I. B. Cohen, *Howard Aiken: Portrait of a Computer Pioneer* (Cambridge, MA: MIT Press, 1999), 212.

11. I will outline here the sense in which an ELS is universal over a class of programs. Consider an example program with five Boolean inputs v_1, v_2, v_3, v_4, v_5 and ten instructions:

$$v_6 = g_6(v_1, v_5),$$

$$L: v_7 = g_7(v_4, v_6),$$

$$\ldots$$

$$v_{15} = g_{15}(v_{11}, v_{14}). \text{ If } (v_{15} = 0) \text{ goto } L.$$

Each of the inputs $v_1, \ldots v_5$ will have a fixed truth value 0 or 1. The ten instructions are executed in order. For example, the first instruction will compute a function g_6 of the values of v_1, v_5 to obtain the 0 or 1 value of variable v_6. The operation g_6 will be from an allowed class. For example, if the OR of two variables is allowed, then g_6 may be that OR function. If the input values are $v_1 = 0, v_2 = 1, v_3 = 1, v_4 = 0, v_5 = 1$, then OR of $v_1 = 0$ and $v_5 = 1$ (as the first instruction "$v_6 = v_1$ OR v_5" demands) will equal 1 since at least one of those two values is 1.

Note that here each g_i depends on v_j for values of j that are smaller than i. Since the values v_i are computed in order of increasing i, the values v_j needed will have been computed earlier and will be available when needed. After the last instruction, if the value of v_{15} is 0, then control will go back to the second instruction, labeled L. Otherwise the program will terminate.

At the highest level, the program is represented in the Mind's Eye by a set of rules. The program variables will be represented by tokens and the values of the variables by attributes that hold for them. The step-by-step updates to the assertions that hold in the Mind's Eye will mimic the values of the variables as the program executes step-by-step.

To realize this, we use special attributes. For each $i = 1, \ldots 15$, we will have two attributes R_i and S_i. For $i = 7$, for example, these will have the following meanings for token a: $R_7(a) = 1$ will assert that token a represents the variable v_7, and $S_7(a) = 1$ will assert that the value of the variable that token a represents is 1. Thus each R_i has value 1 permanently, for exactly one token, and this will be ensured by having management rules NewToken $a, R_i(a_i)$ for $i = 1, \ldots 15$. Then we can represent the value of the variable v_i by the value of $S_i(a)$ for that unique token a for which $R_i(a) = 1$. To set the initial value of each of v_1, v_2, v_3, v_4 and v_5, we have a rule to the effect

$$\text{ForAllTokens } x \left[R_i(x) \Rightarrow S_i(x) = v_i \right]$$

for each of the five values of i. These will ensure that the attribute S_i holds or not for the token that represents v_i according to whether the initial value of v_i is 1 or 0. To mimic the execution of instruction $v_6 = g_6(v_1, v_5)$ a first try is the rule:

$$\text{ForAllTokens } x_6 \text{ ForAllTokens } x_1 \text{ ForAllTokens } x_5$$

$$\left[R_1(x_1) \,\&\, R_5(x_5) \,\&\, g_6(S_1(x_1), S_5(x_5)) \,\&\, R_6(x_6) \Rightarrow S_6(x_6) \right].$$

This identifies the tokens that represent the arguments v_1, v_5 of g_6, determines the value of g_6 implied by the current values of these arguments, and sets the value S_6 of the token that represents v_6 to that value of g_6. This is not quite enough since at each time there is just one instruction in the program to execute, and our rules need to keep track of which one that is to be. To achieve this, for $i = 6, \ldots 15$, we will have two more special attributes I_i and J_i to represent which program instruction is to be executed next. For each i there will be a unique token a for which $I_i(a) = 1$ is set permanently. For that token a, $J_i(a) = 1$ when instruction i is the next instruction to be executed, and $J_i(a) = 0$ otherwise. To indicate that 6 is the first instruction to be executed, one needs to initialize $J_6(a) = 1$ for the a for which $I_6(a) = 1$. The rule given above to mimic $v_6 = g_6(v_1, v_5)$ should be executed only when that is the instruction to be executed. Hence the rule we gave needs to be augmented by inserting further conditions $I_6(x) \,\&\, J_6(x)$ in the left-hand side, with the further qualifier ForAllTokens x in front of the […] parentheses. Also, we need to amplify the rules to update the $J(a)$ values. For Turing completeness, it should be possible to make the choice of next instruction dependent on the values of the program variables, which the goto statement makes possible. To simulate "if $(v_{15} = 0)$ goto L" where L is the instruction for v_7, we would need a rule to make sure that after this instruction is executed, for the appropriate tokens a and b, $J_{15}(a) = 0$ and $J_7(b) = 1$, so that the next instruction will be 7 rather than 15. Each rule here makes several updates simultaneously, to keep track of the values of the variables and the next instruction.

Note that Robust Logic can support more general reasoning mechanisms than chaining. This in turn may support some tasks more efficiently: B. Juba, "Implicit Learning of Common Sense for Reasoning," *International Joint Conference on Artificial Intelligence* (2013): 939–46.

Chapter 10

1. Arrian of Nicomedia, *Anabasis of Alexander*, trans. E. J. Chinnock (London: Hodder and Stoughton, 1884), book 7, chap. 28.

2. Plutarch, *The Life of Alexander*, chap. 6.

3. T. K. Hensch, "Critical Period Regulation," *Annual Review of Neuroscience* 27 (2004): 549–79; J. K. Hartshorne, J. B. Tenenbaum, and S. Pinker, "A Critical Period for Second Language Acquisition: Evidence from 2/3 Million English Speakers," *Cognition* 177 (2018): 263–77.

4. The word "educability" has been used in the psychology literature to refer to the binary question of whether an individual will cope in the conventional education system of the day rather than need special help. Some have proposed to test for this for the purpose of screening. H. Schucman, "Evaluating the Educability of the Severely Mentally Retarded Child," *Psychological Monographs: General and Applied* 74, no. 14 (1960): 1–32. The notion of educability in this book is different.

5. H. L. Roediger and A. C. Butler, "The Critical Role of Retrieval Practice in Long-Term Retention," *Trends in Cognitive Sciences* 15 (2011): 20–27.

Chapter 11

1. Samuel Butler, *Darwin Among the Machines*, to the Editor of the *Press*, Christchurch, New Zealand, June 13, 1863. Also in *Erewhon*, 1872.

2. A. M. Turing, "Intelligent Machinery" (1948), in B. J. Copeland, *The Essential Turing* (Oxford: Oxford University Press, 2004), 410–32; A. M. Turing, "Computing Machinery and Intelligence," *Mind* 59 (1950): 433–60.

3. A. M. Turing, "Can Digital Computers Think?," BBC Radio broadcast, May 15, 1951 (repeated July 3, 1951), in *The Essential Turing*, ed. J. Copeland (Oxford: Oxford University Press, 2004), 486.

4. A. Krizhevsky, I. Sutskever, and G. Hinton, "ImageNet Classification with Deep Convolutional Neural Nets," *Advances in Neural Information Processing Systems*

25 (2012); Y. LeCun, Y. Bengio, and G. Hinton, "Deep Learning," *Nature* 521 (2015): 436–44.

5. J. Deng et al., "ImageNet: A Large-Scale Hierarchical Image Database," *IEEE Conference on Computer Vision and Pattern Recognition*, 2009, https://doi.org/10.1109/CVPR.2009.5206848.

6. E. Korot et al., "Predicting Sex from Retinal Fundus Photographs Using Automated Deep Learning," *Scientific Reports* 11, no. 10286 (2021).

7. A. Buetti-Dinh et al., "Deep Neural Networks Outperform Human Expert's Capacity in Characterizing Bioleaching Bacterial Biofilm Composition," *Biotechnology Reports* 22, no. e00321 (2019).

8. Valiant, *Probably Approximately Correct*.

9. GPT-4 Technical Report, https://arxiv.org/pdf/2303.08774.pdf.

10. A. Vaswani et al., *Attention Is All You Need* (2017), https://arXiv:1706.03762.03762.

11. It is sometimes claimed that deep learning is unique as a learning algorithm in that the number of parameters of the learned classifier can far exceed the number of training examples. Averaging, however, is a long-known process where the number of parameters of a classifier can grow arbitrarily without degrading learning on a fixed dataset. If one learns an arbitrary number of perceptrons, say, on the same dataset, randomizing the order of the examples in each, then averaging the predictions made by the perceptrons on new data will not degrade the predictions as the number of perceptrons goes up. A second setting is the perceptron algorithm itself. The number of examples needed depends on the "margin" of the dataset, namely, how much the positive and negative examples differ, and not on the number of features.

With all practical machine learning techniques, including deep learning, it is often difficult to predict ahead of time how well they will work on a *new* source of data. The degree of success appears to depend on some intrinsic "easiness" in the dataset, such as the margin, which is not evident from just the number of features in the examples.

12. An early experimental attempt to compare pure supervised learning, as in LLMs, with one enhanced with Robust Logic was L. Michael and L. G. Valiant, "A First Experimental Demonstration of Massive Knowledge Infusion," *Proceedings of the 11th International Conference on Principles of Knowledge Representation and Reasoning*, Sydney, Australia, 2008, 378–89.

13. C. Szegedy et al., "Intriguing Properties of Neural Networks," *International Conference on Learning Representations*, 2014.

14. Suppose that a classifier has error rate greater than 10 percent, or, in other words, it is correct on fewer than 90 percent of the examples drawn from the underlying data source D. Then the probability that it is correct on every one of 1,000 training examples *randomly drawn* from D will be less than $(0.9)^{1000}$, which works

out to be less than 10^{-45}, an extremely tiny quantity. Now if the classifier is limited from the outset to be simple, such as one describable by 100 binary bits, then the number of possible classifiers is just 2^{100}, which is fewer than 10^{31}. The probability that *any* among the fewer than 10^{31} such simple classifiers that are less than 90 percent correct on the data source D would accidentally get all the 1,000 random examples correct would then be less than $10^{-45} \times 10^{31}$, which is less than 10^{-14}, still a tiny quantity.

This shows that, except with tiny probability, any 100-bit classifier that agrees with all 1,000 randomly drawn training examples will be a good predictor on future examples, where good means that its error rate is less than 10 percent. This is a rigorous argument that establishes half of the promise of PAC learning, namely, accurate generalization. (This argument does not guarantee efficiency since it says nothing about the process by which one obtains the classifier.) This proof promises accurate generalization with high probability and nothing more.

15. J. Jumper et al., "Highly Accurate Protein Structure Prediction with Alpha-Fold," *Nature* 596 (2021): 583.

16. I. J. Good, "Speculations Concerning the First Ultraintelligent Machine," *Advances in Computers*, vol. 6, (New York: Academic Press, 1965).

17. H. Kissinger, E. Schmidt, and D. Huttenlocher, *The Age of AI: and Our Human Future* (Boston: Little, Brown, 2021).

Chapter 12

1. Harvard University Course Catalog for 2022–23.

2. S. Freeman et al., "Active Learning Increases Student Performance in Science, Engineering, and Mathematics," *PNAS* 111, no. 23 (2014): 8410–15; D. H. Schunk, *Learning Theories: An Educational Perspective*, 6th ed. (London: Pearson Education, 2016); R. E. Slavin, *Educational Psychology*, 13th ed. (London: Pearson Education, 2021).

3. Education Endowment Foundation, "New Evidence-Based Early Years Support," https://educationendowmentfoundation.org.uk, accessed 5/11/2023.

4. Oscar Wilde, *Intentions: The Critic as Artist*, 1891.

5. H. G. Wells, *World Brain* (London: Methuen, 1938).

6. Speech at Launch of Mindset Network, Johannesburg, South Africa, July 16, 2003.

7. Interview with H. G. Wells (September 1937).

8. L. Festinger, *A Theory of Cognitive Dissonance* (Stanford, CA: Stanford University Press, 1957).

9. D. M. Kahan et al., "The Polarizing Impact of Science Literacy and Numeracy on Perceived Climate Change Risks," *Nature Climate Change* 2 (2012): 732–35; D. M. Kahan, "Ideology, Motivated Reasoning, and Cognitive Reflection," *Judgment*

and Decision Making 8, no. 4 (July 2013): 407–24; D. M. Kahan et al., "Cultural Cognition of the Risks and Benefits of Nanotechnology," *Nature Nanotechnology* 4 (2009): 87–91; A. Shtulman, *Scienceblind* (New York: Basic Books, 2017).

10. G. Le Bon, *Psychologie des Foules*, 1895; also Le Bon, *Psychology of Crowds* (London: Sparkling Books, 2009).

11. J. A. F. Stoner, "Risky and Cautious Shifts in Group Decisions: The Influence of Widely Held Values," *Journal of Experimental Social Psychology* 4, no. 4 (1968): 442–59.

12. S. Milgram, *Obedience to Authority: An Experimental View* (London: Tavistock, 1974).

13. P. E. Converse, "The Nature of Belief Systems in Mass Publics (1964)," *Critical Review* 18, no. 1–3 (2006): 1–74.

14. Buddha, *Kālāma Sutta*, Aṅguttara Nikāya, Tika-Nipāta, Sutta no. 65.

15. H. G. Wells, *The Outline of History* (1920), vol. 2, chap. 41, part 4.

16. *Education at a Glance 2022: OECD Indicators*, fig. A1.1, https://www.oecd-ilibrary.org.

17. *Education at a Glance 2022: OECD Indicators*, 246.

18. *Education at a Glance 2022: OECD Indicators*, table D3.1.

Chapter 13

1. J. N. Bremmer, *The Strange World of Human Sacrifice* (Leuven, Belg.: Peeters, 2007); D. Carrasco, *City of Sacrifice* (Boston: Beacon Press, 1999).

2. J. Watts et al., "Ritual Human Sacrifice Promoted and Sustained the Evolution of Stratified Societies," *Nature* 532, no. 10 (2016): 228.

3. "The Report of the Paris Medical Faculty, October 1348," from the website of Martha Carlin, https://sites.uwm.edu/carlin/the-report-of-the-paris-medical-faculty-october-1348/, accessed 5/11/23.

4. V. Narayanamurti and J. Y. Tsao, *The Genesis of Technoscientific Revolutions* (Cambridge, MA: Harvard University Press, 2021).

INDEX

Some of the terms listed are used in the text in a technical sense. For those, the page numbers in italics indicate where the terms are defined.

A NOTE ON THE TYPE

This book has been composed in Arno, an Old-style serif typeface in the classic Venetian tradition, designed by Robert Slimbach at Adobe.